UNITEXT

La Matematica per il 3+2

Volume 144

Editor-in-Chief

Alfio Quarteroni, Politecnico di Milano, Milan, Italy; École Polytechnique Fédérale de Lausanne (EPFL), Lausanne, Switzerland

Series Editors

Luigi Ambrosio, Scuola Normale Superiore, Pisa, Italy

Paolo Biscari, Politecnico di Milano, Milan, Italy

Ciro Ciliberto, Università di Roma "Tor Vergata", Rome, Italy

Camillo De Lellis, Institute for Advanced Study, Princeton, New Jersey, USA

Massimiliano Gubinelli, Hausdorff Center for Mathematics, Rheinische Friedrich-Wilhelms-Universität, Bonn, Germany

Victor Panaretos, Institute of Mathematics, École Polytechnique Fédérale de Lausanne (EPFL), Lausanne, Switzerland

Lorenzo Rosasco, DIBRIS, Università degli Studi di Genova, Genova, Italy; Center for Brains Mind and Machines, Massachusetts Institute of Technology, Cambridge, Massachusetts, US; Istituto Italiano di Tecnologia, Genova, Italy

The **UNITEXT - La Matematica per il 3+2** series is designed for undergraduate and graduate academic courses, and also includes books addressed to PhD students in mathematics, presented at a sufficiently general and advanced level so that the student or scholar interested in a more specific theme would get the necessary background to explore it.

Originally released in Italian, the series now publishes textbooks in English addressed to students in mathematics worldwide.

Some of the most successful books in the series have evolved through several editions, adapting to the evolution of teaching curricula.

Submissions must include at least 3 sample chapters, a table of contents, and a preface outlining the aims and scope of the book, how the book fits in with the current literature, and which courses the book is suitable for.

For any further information, please contact the Editor at Springer: francesca.bonadei@springer.com

THE SERIES IS INDEXED IN SCOPUS
***

UNITEXT is glad to announce a new series of free webinars and interviews handled by the Board members, who rotate in order to interview top experts in their field.

Access this link to subscribe to the events: https://cassyni.com/events/TPQ2UgkCbJvvz5QbkcWXo3

Antoine Chambert-Loir

Information Theory

Three Theorems by Claude Shannon

Antoine Chambert-Loir
UFR de Mathématiques
Université Paris Cité
PARIS CEDEX 13, France

ISSN 2038-5714   ISSN 2532-3318 (electronic)
UNITEXT
ISSN 2038-5722   ISSN 2038-5757 (electronic)
La Matematica per il 3+2
ISBN 978-3-031-21560-5   ISBN 978-3-031-21561-2 (eBook)
https://doi.org/10.1007/978-3-031-21561-2

© The Editor(s) (if applicable) and The Author(s), under exclusive license to Springer Nature Switzerland AG 2022

This work is subject to copyright. All rights are solely and exclusively licensed by the Publisher, whether the whole or part of the material is concerned, specifically the rights of reprinting, reuse of illustrations, recitation, broadcasting, reproduction on microfilms or in any other physical way, and transmission or information storage and retrieval, electronic adaptation, computer software, or by similar or dissimilar methodology now known or hereafter developed.

The use of general descriptive names, registered names, trademarks, service marks, etc. in this publication does not imply, even in the absence of a specific statement, that such names are exempt from the relevant protective laws and regulations and therefore free for general use.

The publisher, the authors, and the editors are safe to assume that the advice and information in this book are believed to be true and accurate at the date of publication. Neither the publisher nor the authors or the editors give a warranty, expressed or implied, with respect to the material contained herein or for any errors or omissions that may have been made. The publisher remains neutral with regard to jurisdictional claims in published maps and institutional affiliations.

This Springer imprint is published by the registered company Springer Nature Switzerland AG
The registered company address is: Gewerbestrasse 11, 6330 Cham, Switzerland

To all contributors of Wikipedia

Preface

The mathematical theory of communication studies in a mathematical way the conditions under which one can transmit data, in particular at which speed, and with what reliability. We owe this expression to Claude SHANNON (1916–2001), and a variant of it gave rise to a scientific field, *information theory*.

Mathematician, electrical engineer, computer scientist, cryptologist, Shannon's scientific activity encompassed all these fields, be it during his PhD Thesis at MIT, within the US Army during the World War II, or at the MIT of which he became a professor in 1956.

One can recognize many sources of inspiration of Shannon's works, such as those of the mathematicians Norbert Wiener and Andrey Kolmogorov, but his direct motivation seems related to the military effort of World War II. In a report for the US National Defense Research Committee devoted to fire-control systems, BLACKMAN ET AL (1946) observe that

> "There is an obvious analogy between the problem of smoothing the data to eliminate or reduce the effect of tracking errors and the problem of separating a signal from interfering noise in communications systems."

Therefore he proposed, in a paper written in 1945 and published a few years later, (SHANNON, 1949b), a theoretical analysis of cryptographic systems, and raised the question of their safety in terms of information theory: if the opponent obtained a given part of the ciphered message, what amount information does he actually hold?

After the war, Shannon widened (or rather *erased*) the context of his ideas and proposed in his foundational paper SHANNON (1948)[1] that a communication system should be modelled by the following diagram (usually written in a row, but the width of these pages made it illegible):

The *source* is the entity that holds some information, some *message* to pass to its *destination*. It may be a radio or television station, a journal, a web site, you and me wishing to tell some worrying news on the phone or by email, a space probe taking

---

[1] The next year, Shannon published this paper as a book, preceded by an introduction SHANNON & WEAVER (1949). Let us note that the title changed slightly, from *a* to *the* mathematical theory of communication.

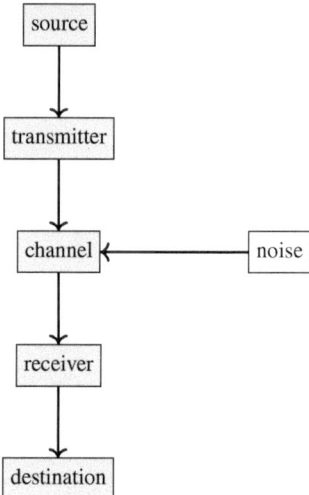

pictures of the planets it meets, etc. The message can be a text, a photograph, a piece of music, or any combination of these...

The *emitter* is the physical device by which this message is emitted, such as a radio or TV emitter. At the time of Shannon, transmission was often analogue. For radio or telephone, sound is represented by the amplitude of air pressure on a microphone, which the latter transforms into an electric signal. The question is then to pass this amplitude, as a function of time, or by two such functions for a stereo sound signal. For color TV, one needs to transmit three color amplitudes (red/green/blue) at each point of the screen, and two amplitudes for the sound, all of them depending on time. Nowadays, signals are mostly numeric: the message is transformed into a sequence of numbers, which then are to be transmitted. This holds for television, telephone, Internet..., with the notable exception of FM radio, since Digital Audio Broadcasting is barely taking off.

The *receiver* is the device by which the addressee receives this message: the radio or television, possibly linked to some decoder in the case of digital television, a telephone, a computer linked to the Internet network, etc.

Between emitter and receiver, what is passed is not exactly the initial message, but its conversion to a *signal*, and the *channel* is the material by which this signal is transmitted from emitter to receiver: air for the analogue radio/TV transmission, the optic fiber of the Internet provider, the copper wires of the telephone network, etc. As with every physical device, this channel is subject to perturbations—*noise*— so that the signal that is received differs from what was sent by the emitter, hence the destination receives a different message than what the source had intended to send.

The mathematical theory of communication aims at analysing under which conditions a given channel, subject to some noise, may, or may not, transmit a given message. Two theorems of SHANNON (1948) answer the following questions:

1. At what speed can one transmit a message?
2. In the presence of noise, can one transmit a message in a reliable way?

If the channel allows c symbols to pass per second it would seem obvious that one needs N/c seconds to pass an N-symbol message, but can one do better? And if some symbols have been wrongly transmitted, can one eventually correct them?

The basic assumption of the theory lies in the fact that the messages to be transmitted, because of their origin, are not arbitrary. If it is a text, certain letters will be more frequent than the other; if it is the recording of a voice, or of a musical piece, certain sound frequencies will not be present in the signal, not even taking into account that the loudspeakers or the headphones cannot reproduce them, nor the human ear perceive them.

In his paper, SHANNON (1948) proposes a definition of the "amount of information" contained in a message, which he calls *entropy*. More precisely, this is the amount of information contained in all signals susceptible to being transmitted. Its definition uses probability theory and we study it in chapter 1. We define the entropy of a random variable and some variants:

- Conditional entropy, representing the additional amount of information that a random variable gives, with respect to some other;
- Mutual information between two random variables, representing in a symmetric way the information that each gives about the other;
- The entropy rate of a sequence of random variables (aka, a *stochastic process*), representing the mean information brought by each of them.

In particular, we compute the entropy rate in the important case of *Markov processes*, for which each random variable is independent of all preceding ones.

Since this chapter is of a quite abstract nature, we had to precede it with a chapter that "recalls" the basic facts of probability theory.

Chapter 2 is devoted to *coding*, that is to the two theorems of Shannon that we evoked earlier, and which allow us to analyse two aspects of signal transmission: possibility of compression, and possibility of correcting transmission errors. These two facts lie at the foundations of the modern theory of digital telecommunications.

The next chapter discusses the question of *sampling*: this is the first phase of digitization of some signal, where it is transformed into a sequence of numbers. The "sampling theorem" is usually attributed to Shannon, but in his paper (SHANNON, 1949a), he viewed it as "common knowledge in the communication art." It gives hypotheses that ensure that sampling can be done without losing any information. The main mathematical tool of this chapter is Fourier theory, which allows the different frequencies appearing in a signal to be separated. A significant part of the chapter is devoted to establishing the main results of the theory of Fourier series, and of the Fourier transform.

Each chapter is accompanied by many exercises, whose solutions are given in the final chapter.

I added some historical notes to embody the names of the mathematicians or physicists met throughout the book. Not as rigorous as they should be, these short notes are an invitation to embark on a reading of these works that would not neglect the historic, social, scientific conditions of their realizations, be it, for example, about Joseph Fourier, professor at École polytechnique, hired by Napoléon as scientific advisor of the large contingent of scientists that accompanied his military expedition to Egypt, or about Claude Shannon, whose works during World War II aimed at resisting the jamming of electric communication.

Writing these notes made me more aware of the way the mathematical and physical sciences, information theory in particular, are intimately linked with military efforts, more generally to the military-industrial complex. At a time of history where mathematics and computer science play an important role, by their handling of enormous amounts of digital data or encryption of communications, to the point that one can talk of "surveillance capitalism", if that awareness reaches the readers of this book then this meager effort shall not have been made in vain.

The English literature already includes many introductory books about information theory. Written by mathematicians, engineers, computer scientists that are more competent that I am regarding this topic, these books are often excellent, more comprehensive, and often reach much more advanced results. In particular, the book of COVER & THOMAS (2006) has been very useful to me, from its overall presentation to some of the exercises; readers who can read this book with profit will recognize my debt to it. Nevertheless, I hope that the present book, of limited size and ambition, will serve as a pleasant introduction to this beautiful topic.

I taught this course for three years at Université Paris Cité (formerly known as Paris-7), within the Maths and Computer Science master program. I thank Georges Skandalis and Justin Salez for having passed me their notes and lists of exercises that they used for this course. I also thank Guillaume Garrigos for having thoroughly worked out the exercises sessions, and for having insisted that my arguments often needed more details.

Colleague and publisher Rached Mneimné welcomed the French edition of this little book. With Alain Debreil, he proof-read that version in full detail, and Alain provided a few more diagrams. Thanks to them!

I finally thank the students of the master program for their participation, in particular for those who followed this course during the academic year 2020–2021, when the COVID-19 pandemic constrained them to remote study.

Contents

Preface vii

0. Some bits of probability theory 1
    0.1. Summable families 1
    0.2. Probability theory 9
    0.3. Discrete random variables 11
    0.4. Independence, conditional expectation 17
    Exercises 20

1. Entropy and mutual information 23
    1.1. Entropy of a discrete random variable 23
    1.2. Conditional entropy 28
    1.3. Mutual information 32
    1.4. Entropy rate 38
    1.5. Entropy rate of Markov processes 40
    Exercises 49

2. Coding 55
    2.1. Codes 56
    2.2. The Kraft–McMillan inequality 58
    2.3. Optimal codes 61
    2.4. The law of large numbers, and compression 67
    2.5. Transmission capacity of a channel 74
    2.6. Coding adapted to a transmission channel 80
    Exercises 89

3. Sampling 95
    3.1. Continuous signals and discrete signals 96
    3.2. The Fourier series of a periodic function 98
    3.3. The main theorems of the theory of Fourier series 103
    3.4. Convolution and Dirichlet's theorem 111

	3.5. Fourier transformation 119
	3.6. The sampling theorem 123
	3.7. The uncertainty principle in communication theory 126
	Exercises 132

Notation 137

4. **Solutions of the exercises** 139
   - 4.0. Some bits of probability theory 139
   - 4.1. Entropy and mutual information 148
   - 4.2. Coding 171
   - 4.3. Sampling 189

References 205

Index 207

Chapter 0.
Some bits of probability theory

This chapter aims at recalling some definitions and some elementary results in the field of discrete probability theory.

We owe to KOLMOGOROV (1956)[1] (the first edition, in German, dates from 1933) the foundations of the theory of probability on the grounds of the Lebesgue integral (see footnote, page 98). The discrete random variables that we consider in this little book do not require such an apparatus and, in most cases, the computations will follow from manipulations of finite sums. In some cases, the discrete random variables that we consider might take infinitely many values, and the finite sums must then be replaced by series, or rather summable families, because their index set is not naturally ordered. The first section recalls this theory; it might be wise not to read it too thoroughly.

We then give a formal definition, maybe too formal in fact, of a probabilistic space. To avoid having to bother with possible elements of null probability, I have been led to assume that the universe is "complete" for the considered probability.

The next section recalls the definitions of (discrete) random variables on a probabilistic space, of their law, their expectation and their variance (provided they exist).

We finally recall the definition of independence, a crucial notion in probability theory. It is rather elementary when it comes to independence of two events, but becomes more subtle when applied to random variables. We end this section with the notions of conditional expectation and variance.

0.1. Summable families

We may have to handle infinite sums indexed by a set which is neither the set of integers, nor in obvious bijection with the set of integers. The theory of summable

---

[1] Andrey KOLMOGOROV (1903–1987), Soviet mathematician whose fundamental works encompassed probability theory, turbulence in fluid mechanics, celestial mechanics, complexity theory, statistics, topology...

families describes in which context such sums exist, and gives methods to compute them.

Definition 0.1.1. — *Let A be a set, and let $(z_a)_{a\in A}$ be a family of complex numbers indexed by A. One says that this family is* summable *if there exists a complex number z such that, for every $\varepsilon > 0$, there is a finite subset B of A such that*

$$\left| z - \sum_{a\in C} z_a \right| \leq \varepsilon,$$

for every finite subset C of A such that $B \subseteq C$. One then says that (z_a) is summable with sum z.

If the set A is finite, then the family $(z_a)_{a\in A}$ is summable with sum the sum $z = \sum_{a\in A}$ of this finite family: in the definition, it suffices to take $B = A$.

Lemma 0.1.2. — *Let $(z_a)_{a\in A}$ be a family indexed by a set A and let z, z' be complex numbers. Assume that the family (z_a) is summable with sum z and is summable with sum z'. Then $z = z'$.*

Given this lemma, it is legitimate to call the *sum* of a summable family (z_a) the unique complex number z such that this family is summable with sum z; we denote it by $\sum_{a\in A} z_a$.

The whole theory of summable families aims at giving means to compute this sum, in particular to give workable conditions under which it "behaves as" a finite sum.

Proof. — Let ε be a strictly positive real number. Let B be a finite subset of A such that $|z - \sum_{a\in C} z_a| \leq \varepsilon$ for every finite subset C of A that contains B. Let us choose B' in a similar manner with respect to z' and let us set $C = B \cup B'$. Then $|z - \sum_{a\in C} z_a| \leq \varepsilon$ and $|z' - \sum_{a\in C} z_a| \leq \varepsilon$, so that $|z - z'| \leq 2\varepsilon$. Since ε is arbitrary, this implies $z = z'$. □

Lemma 0.1.3. — *For a family $(z_a)_{a\in A}$ of positive real numbers to be summable, it is necessary and sufficient that the sums $\sum_{a\in B} z_a$ be bounded from above, when B runs among all finite subsets of A. Then we have*

$$\sum_{a\in A} z_a = \sup_{\substack{B\subseteq A \\ B \text{ is finite}}} \left(\sum_{a\in B} z_a \right).$$

When these sums are not bounded from above, one sets $\sum_{a\in A} z_a = +\infty$.

Proof. — Let us first assume that the family $(z_a)_{a\in A}$ is summable with sum z. Let B be a finite subset of A such that $|z - \sum_{a\in C} z_a| \leq 1$ for every finite subset C of A that contains B. For every finite subset C of A, we then have

$$\sum_{a\in C} z_a \leq \sum_{a\in B\cup C} z_a \leq |z| + 1,$$

so that the set of considered sums are bounded from above.

0.1. Summable families.

Conversely, let us assume that these sums are bounded from above, and let z be their least upper bound; let us prove that (z_a) is summable with sum z. Let $\varepsilon > 0$, and let B be a finite subset of A such that $z - \varepsilon < \sum_{a \in B} z_a \leq z$. Let C be a finite subset of A such that $B \subseteq C$. One has $\sum_{a \in C} z_a \leq z$, by definition of z. On the other hand, since $z_a \geq 0$ for every $a \in A$, we have

$$\sum_{a \in C} z_a = \sum_{b \in B} z_b + \sum_{a \in C-B} z_a \geq \sum_{b \in B} z_b \geq z - \varepsilon.$$

Consequently, $|z - \sum_{a \in C} z_a| \leq \varepsilon$. This proves that the family $(z_a)_{a \in A}$ is summable with sum z. □

Properties 0.1.4. — a) *If the families $(z_a)_{a \in A}$ and $(z'_a)_{a \in A}$ are summable, then so is the family $(z_a + z'_a)_{a \in A}$, and one has*

$$\sum_{a \in A} (z_a + z'_a) = \sum_{a \in A} z_a + \sum_{a \in A} z'_a.$$

Let $z = \sum_{a \in A} z_a$ and $z' = \sum_{a \in A} z'_a$. Let $\varepsilon > 0$, and let B be a finite subset of A such that $|z - \sum_{a \in C} z_a| < \varepsilon/2$ for every finite subset C of A such that $B \subseteq C$. Similarly, let B' be a finite subset of A such that $|z' - \sum_{a \in C} z'_a| < \varepsilon/2$ for every finite subset C of A such that $B' \subseteq C$. The set $B \cup B'$ is finite; moreover, for every finite subset C of A such that $B \cup B' \subseteq C$, one has

$$\left| (z + z') - \sum_{a \in C} (z_a + z'_a) \right| \leq \left| z - \sum_{a \in C} z_a \right| + \left| z' - \sum_{a \in C} z'_a \right|$$
$$< \varepsilon/2 + \varepsilon/2 = \varepsilon.$$

This proves that the family $(z_a + z'_a)_{a \in A}$ is summable with sum $z + z'$.

b) *Let λ be a complex number. If a family $(z_a)_{a \in A}$ is summable, then so is the family $(\lambda z_a)_{a \in A}$, and one has*

$$\sum_{a \in A} \lambda z_a = \lambda \sum_{a \in A} z_a.$$

Let $\varepsilon > 0$ and let $z = \sum_{a \in A} z_a$. Let B be a finite subset of A such that for every finite subset C of A such that $B \subseteq C$, one has $|z - \sum_{a \in C} z_a| < \varepsilon/(1 + |\lambda|)$. Then for every such subset C, one has

$$\left| \lambda z - \sum_{a \in C} \lambda z_a \right| = |\lambda| \times \left| z - \sum_{a \in C} z_a \right| < \varepsilon \frac{|\lambda|}{1 + |\lambda|} < \varepsilon,$$

which proves that the family $(\lambda z_a)_{a \in A}$ is summable and that its sum is equal to λz.

c) *If a family $(z_a)_{a \in A}$ is summable, then so is the family $(\overline{z_a})_{a \in A}$, and one has*

$$\sum_{a \in A} \overline{z_a} = \overline{\sum_{a \in A} z_a}.$$

Let $\varepsilon > 0$ and let $z = \sum_{a \in A} z_a$. Let B be a finite subset of A such that for every finite subset C of A such that $B \subseteq C$, one has $|z - \sum_{a \in C} z_a| < \varepsilon$. For every such subset C, we then have

$$\left| \overline{z} - \sum_{a \in C} \overline{z_a} \right| = \left| z - \sum_{a \in C} z_a \right| < \varepsilon,$$

which proves that the family $(\overline{z_a})_{a \in A}$ is summable with sum $\overline{z}$.

Proposition 0.1.5. — *Let $(z_a)_{a \in A}$ be a family of complex numbers.*

a) *(Cauchy[2] criterion) The family $(z_a)_{a \in A}$ is summable if, and only if, for every $\varepsilon > 0$, there exists a finite subset B of A such that one has $\left| \sum_{a \in C} z_a \right| \leq \varepsilon$ for every finite subset C of A which is disjoint from B.*
b) *The family $(z_a)_{a \in A}$ is summable if, and only if, there exists a real number M such that $\left| \sum_{a \in B} z_a \right| \leq M$ for every finite subset B of A.*

Proof. — a) The necessity of this criterion is absolutely parallel to that of the Cauchy criterion for series. Let us assume that the family (z_a) is summable with sum z. Let $\varepsilon > 0$; let us choose a finite subset B of A such that $|z - \sum_{a \in C} z_a| \leq \varepsilon/2$ for every finite subset C of A that contains B. Let then C be a finite subset of A which is disjoint from B; we have

$$\left| \sum_{a \in C} z_a \right| = \left| \sum_{a \in B \cup C} z_a - \sum_{a \in B} z_a \right|$$

$$\leq \left| \sum_{a \in B \cup C} z_a - z \right| + \left| \sum_{a \in B} z_a - z \right|$$

$$\leq 2\varepsilon/2 = \varepsilon.$$

To prove that this condition suffices, let us define by induction an increasing sequence (B_n) of finite subsets of A by first setting $B_0 = \emptyset$; then, for any $n \geq 1$ such that B_{n-1} has been defined, let us choose for B_n some finite subset of A containing B_{n-1} such that $|\sum_{a \in C} z_a| \leq 1/n$ for every finite subset C of A which is disjoint from B_n. For any integer n, let us then set $u_n = \sum_{a \in B_n} z_a$. If m and n are integers such that $n \geq m \geq 1$, we have

---

[2] Augustin Louis CAUCHY (1789–1857) was a French mathematician, with very important works in real and complex analysis as well as in algebra. Although his arguments lacked rigor, we owe him the first precise definitions of the convergence of a series, of limits and differentiability, and the statement of the intermediate value theorems. He also worked in celestial mechanics. His negligence regarding memoirs that Abel and Galois had submitted to the French Académie des sciences certainly led to some delay in the dissemination of their discoveries.

0.1. Summable families.

$$|u_m - u_n| = \left|\sum_{a\in B_n} z_a - \sum_{a\in B_m} z_a\right| = \left|\sum_{a\in B_n - B_m} z_a\right| \leq \frac{1}{m}.$$

Consequently, the sequence (u_n) of complex numbers satisfies the Cauchy convergence criterion, hence converges to some complex number u in $\mathbf{C}$. Let m be an integer; when n goes to infinity in the previous inequality, we obtain the inequality $|u_m - u| \leq 1/m$.

Let us now prove that the family (z_a) is summable with sum z. Let $\varepsilon > 0$, and let m be an integer such that $\frac{2}{m} < \varepsilon$. Let C be a finite subset of A containing B_m; we have

$$\left|u - \sum_{a\in C} z_a\right| = \left|(u - u_m) - \sum_{a\in C-B_m} z_a\right|$$

$$\leq |u - u_m| + \left|\sum_{a\in C-B_m} z_a\right|$$

$$\leq \frac{1}{m} + \frac{1}{m} \leq \varepsilon.$$

This proves the desired result.

b) The condition is necessary. Indeed, let us assume that the family $(z_a)_{a\in A}$ is summable with sum z, and let C_1 be a finite subset of A such that $|z - \sum_{a\in C} z_a| < 1$ for every finite subset C of A such that $C_1 \subseteq C$. Let then B be a finite subset of A. The set $B \cup C_1$ is finite, and we have

$$\sum_{a\in B} z_a = z - z + \sum_{a\in B\cup C_1} z_a - \sum_{a\in C_1-B} z_a,$$

so that

$$\left|\sum_{a\in B} z_a\right| \leq |z| + \left|z - \sum_{a\in B\cup C_1} z_a\right| + \sum_{a\in C_1-B} |z_a|$$

$$\leq |z| + 1 + \sum_{a\in C_1} |z_a|.$$

This proves the desired upper bound, with $M = |z| + 1 + \sum_{a\in C_1} |z_a|$.

Conversely, let us prove that this condition is sufficient. We first consider the case where the family $(z_a)_{a\in A}$ only takes positive real values. By assumption, the family $\left(\sum_{a\in B} z_a\right)_B$, indexed by the set of all finite subsets B of A, has an upper bound; by lemma 0.1.3, the family $(z_a)_{a\in A}$ is summable.

Let us now assume that the family (z_a) takes real values. For every $a \in A$, set $z_a^+ = \sup(z_a, 0)$ and $z_a^- = \sup(-z_a, 0)$, so that $z_a = z_a^+ - z_a^-$. The two families $(z_a^+)_{a\in A}$ and $(z_a^-)_{a\in A}$ have positive terms. Let us check that they satisfy the condition of the statement. Let $M > 0$ be such that for any finite subset B of A, one has $|\sum_{a\in B} z_a| \leq M$. Let B be a finite subset of A; let B^+ be the set of $a \in A$ such that $z_a \geq 0$. By assumption, one has $|\sum_{a\in B^+} z_a| \leq M$; since $z_a^+ = z_a$ for

$a \in B^+$ and $z_a^+ = 0$ otherwise, we then have

$$\left|\sum_{a \in B} z_a^+\right| = \left|\sum_{a \in B^+} z_a\right| \leq M.$$

One proves in a similar way that $\left|\sum_{a \in B} z_a^-\right| \leq M$. By the initial case, the two families $(z_a^+)_{a \in A}$ and $(z_a^-)_{a \in A}$ are summable, hence so is the family $(z_a)_{a \in A}$. Let us finally prove the general case. For any $a \in A$, set $x_a = \text{Re}(z_a)$ and $y_a = \text{Im}(z_a)$, so that $z_a = x_a + i y_a$. Let M be a real number such that $\left|\sum_{a \in B} z_a\right| \leq M$ for every finite subset B of A. For every such B, we then have

$$\left|\sum_{a \in B} x_a\right| \leq \left|\sum_{a \in B} x_a + i \sum_{a \in B} y_a\right| \leq M,$$

and

$$\left|\sum_{a \in B} y_a\right| \leq \left|\sum_{a \in B} x_a + i \sum_{a \in B} y_a\right| \leq M.$$

This implies that the families $(x_a)_{a \in A}$ and $(y_a)_{a \in A}$ satisfy the condition of the statement. Since they take real values, the preceding case implies that they are summable. Then the family $(z_a)_{a \in A}$ is summable too. □

Corollary 0.1.6. — *Let $(z_a)_{a \in A}$ be a summable family, and let B be a subset of A. The family $(z_a)_{a \in B}$ is then summable.*

Let us moreover assume that the family (z_a) has real positive terms. Then one has

$$\sum_{a \in B} z_a \leq \sum_{a \in A} z_a.$$

Proof. — The first assertion results from any one of the criteria of proposition 0.1.5. To prove the second, let us recall that $\sum_{a \in B} z_a$ is the upper bound, for all finite subsets B_1 of B, of the sums $\sum_{a \in B_1} z_a$, while $\sum_{a \in A} z_a$ is the upper bound, for all finite subsets A_1 of A, of the sums $\sum_{a \in A_1} z_a$. One then has $\sum_{a \in B_1} z_a \leq \sum_{a \in A} z_a$ for every finite subset B_1 of B. Then $\sum_{a \in B} z_a \leq \sum_{a \in A} z_a$. □

Remark 0.1.7. — Let $(z_a)_{a \in A}$ be a summable family of complex numbers. Let n be a natural integer such that $n \geq 1$ and apply the Cauchy criterion with $\varepsilon = 1/n$. There exists a finite subset B_n of A such that, for every $a \in A - B_n$, one has $|z_a| \leq 1/n$. Let B be the union of all B_n; this is a countable subset of A. For $a \in A - B$, one has $|z_a| \leq 1/n$ for all n, hence $z_a = 0$. In other words, the *support* of the family (z_a), that is the set of all $a \in A$ such that $z_a \neq a$, is countable.

Proposition 0.1.8. — *Let A be a set and let $(z_a)_{a \in A}$ be a summable family of complex numbers.*

a) Let $(A_1, \ldots, A_n)$ be a finite partition of the set A. For every $j \in \{1, \ldots, n\}$, the family $(z_a)_{a \in A_j}$ is summable. Moreover, one has

0.1. Summable families.

$$\sum_{j=1}^{n}\left(\sum_{a\in A_j} z_a\right) = \sum_{a\in A} z_a.$$

b) *Let $\varphi: A \to B$ be a map. For every $b \in B$, the family $(z_a)_{a\in\varphi^{-1}(b)}$ is summable; let u_b be its sum. The family $(u_b)_{b\in B}$ is summable and its sum is equal to $\sum_{a\in A} z_a$. In other words, we have*

$$\sum_{b\in B}\left(\sum_{a\in\varphi^{-1}(b)} z_a\right) = \sum_{a\in A} z_a.$$

With the usual conventions for sums involving $+\infty$, the analogous assertions hold for any family (z_a) with positive terms.

Assertion *a)* is a particular case of assertion *b)*, applied with $B = \{1, \ldots, n\}$ and φ the map that identifies j and A_j. However, we have to prove it as an intermediate step.

Proof. — We first treat the case of summable families. By corollary 0.1.6, the families $(z_a)_{a\in A_j}$ (for $j \in \{1, \ldots, n\}$) and $(z_a)_{a\in\varphi^{-1}(b)}$ (for $b \in B$) are summable.

a) For $j \in \{1, \ldots, n\}$, set $u_j = \sum_{a\in A_j} z_a$. Let $\varepsilon > 0$, and let ε' be a real number such that $n\varepsilon' \leq \varepsilon$. For every $j \in \{1, \ldots, n\}$, there exists, by definition of u_j, a finite subset B_j of A_j such that for every finite subset C_j of A_j such that $B_j \subseteq C_j$, we have $|u_j - \sum_{a\in C_j}| \leq \varepsilon'$. Let B be the union of these B_j, for $j \in \{1, \ldots, n\}$, and let C be a finite subset of A containing B; for $j \in \{1, \ldots, n\}$, set $C_j = C \cap A_j$, so that $(C_j)_{1\leq j\leq n}$ is a partition of C and $B_j \subseteq C_j$ for all j.
Then,

$$\left|\sum_{j=1}^{n} u_j - \sum_{a\in C} z_a\right| = \left|\sum_{j=1}^{n}\left(u_j - \sum_{a\in C_j} z_a\right)\right|$$
$$\leq \sum_{j=1}^{n}\left|u_j - \sum_{a\in C_j} z_a\right| \leq n\varepsilon' \leq \varepsilon.$$

This implies the desired equality.

b) Let us prove that the family $(u_b)_{b\in B}$ is summable with sum $z = \sum_{a\in A} z_a$. Let $\varepsilon > 0$. By the definition of z, there exists a finite subset A_1 of A such that $|z - \sum_{a\in C} z_a| \leq \varepsilon/2$ for every finite subset C of A that contains A_1. As in the proof of the Cauchy criterion, it follows that for every subset A' of A, finite or not, that is disjoint from A_1, one has $|\sum_{a\in A'} z_a| \leq \varepsilon$. Indeed, set $z' = \sum_{a\in A'} z_a$. Let $\delta > 0$, and let A'_1 be a finite subset of A' such that $|z' - \sum_{a\in C'} z_a| \leq \delta$ for every finite subset C' of A' such that $A'_1 \subseteq C'$. Writing

$$z' = \left(z' - \sum_{a\in C'} z_a\right) + \left(z - \sum_{a\in A_1} z_a\right) - \left(z - \sum_{a\in A_1\cup C'} z_a\right),$$

we see that $|z'| \leq \delta + \varepsilon$. Since δ is arbitrary, we have $|z'| \leq \varepsilon$.

Let $B_1 = \varphi(A_1)$, and let C be a finite subset of A containing B_1. By the first case, applied to the partition $(\varphi^{-1}(b))_{b \in C}$ of $\varphi^{-1}(C)$, we have the equality

$$\sum_{b \in C} u_b = \sum_{a \in \varphi^{-1}(C)} z_a.$$

Applying the first case to the partition $(\varphi^{-1}(C), A - \varphi^{-1}(C))$ of A, we observe the equality

$$z = \sum_{a \in A} z_a = \sum_{a \in \varphi^{-1}(C)} z_a + \sum_{a \in A - \varphi^{-1}(C)} z_a,$$

from which we conclude that

$$z - \sum_{b \in C} u_b = \sum_{a \in A - \varphi^{-1}(C)} z_a.$$

On the other hand, for any $a \in A_1$, we have $\varphi(a) \in B_1$, hence $\varphi(a) \in C$, which proves that $A_1 \subseteq \varphi^{-1}(C)$; equivalently, $A - \varphi^{-1}(C)$ is disjoint from A_1. Consequently, the absolute value of the right-hand side of the preceding equality is smaller than ε; we thus have $|z - \sum_{b \in C} u_b| \leq \varepsilon$. This proves that the family $(u_b)_{b \in B}$ is summable with sum z.

Let us now treat the case of families with positive terms. Assertion *a)* is obvious if there exists a j such that $(z_a)_{a \in A_j}$ is not summable: both sides of the equality are equal to $+\infty$. Conversely, let us assume that $(z_a)_{a \in A_j}$ is summable, for every j. Let C be a finite subset of A; for every j, set $C_j = C \cap A_j$. Then

$$\sum_{a \in C} z_a = \sum_{j=1}^{n} \sum_{a \in C_j} z_a \leq \sum_{j=1}^{n} \left(\sum_{a \in A_j} z_a \right).$$

This proves that the family $(z_a)_{a \in A}$ is summable, and we conclude by the case that has been already treated.

To prove assertion *b)*, let us assume that the left-hand side is finite. For every finite subset C of A, we have

$$\sum_{a \in C} z_a \leq \sum_{b \in B} \sum_{a \in \varphi^{-1}(b) \cap C} z_a \leq \sum_{b \in B} \left(\sum_{a \in \varphi^{-1}(b)} z_a \right).$$

This proves that the family $(z_a)_{a \in A}$ is summable. Conversely, let us assume that the left-hand side is infinite and let us prove that the same holds for the right-hand side. This is obvious if there exists a $b \in B$ such that $\sum_{a \in \varphi^{-1}(b)} z_a$ is infinite. Suppose this does not happen and let M be a positive real number. There exists a finite subset B_1 of B such that

$$\sum_{b \in B_1} \sum_{a \in \varphi^{-1}(B_1)} z_a \geq M + 1.$$

Let $\varepsilon = 1/\sup(1, \operatorname{Card}(B_1))$. For every $b \in B_1$, there exists a finite subset C_b of $\varphi^{-1}(b)$ such that $\sum_{a \in A_b} z_a \geq \sum_{a \in \varphi^{-1}(B_1)} -\varepsilon$; let C be the union of the family (C_b).

We thus have

$$\sum_{a \in C} z_a = \sum_{b \in B_1} \sum_{a \in C_b} z_a \geq \sum_{b \in B_1} \left(-\varepsilon + \sum_{a \in C_b} z_a \right) \geq -1 + M + 1 = M.$$

This proves that the sum of the family $(z_a)_{a \in A}$ is $+\infty$. □

0.2. Probability theory

0.2.1. — Nowadays, probability theory is formalized within the framework of measure theory. We consider (let us pretend it is fixed once and for all) a set Ω, a set $\mathscr{E}$ of subsets of Ω and a map $\mathbf{P}\colon \mathscr{E} \to [0; 1]$. The set Ω is called the *sample space*, or the *universe*, and the elements of $\mathscr{E}$ are called *events*; if A is an event, $\mathbf{P}(A)$ is its *probability*. The set $\mathscr{E}$ of events and the probability function $\mathbf{P}$ are assumed to satisfy the following axioms.

(P_1) One has $\Omega \in \mathscr{E}$.
(P_2) The union $\bigcup_{n \in \mathbf{N}} A_n$ of a sequence (A_n) of events is again an event.
(P_3) The complement $\Omega - A$ of an event A is an event.

Because of the formula

$$\bigcap_{n \in \mathbf{N}} A_n = \Omega - \bigcup_{n \in \mathbf{N}} (\Omega - A_n),$$

the intersection $\bigcap_{n \in \mathbf{N}} A_n$ of a sequence of events is again an event. Applying these properties to a sequence that has only two distinct terms A and B, we also deduce that if A and B are events, then $A \cup B$ and $A \cap B$ are events. More generally, the union of a *countable* family of events (meaning, its index set is in bijection with a subset of $\mathbf{N}$) and the intersection of a countable family of events are events too.

We also see that $\emptyset = \Omega - \Omega$ is an event.

These first three axioms can thus be summarized by saying that $\mathscr{E}$ is a σ-*algebra* on the sample space Ω.

The three additional axioms concern the probability function $\mathbf{P}$ and state that $\mathbf{P}$ is a positive *measure* of total mass 1 for which the σ-algebra $\mathscr{E}$ is complete.

(P_4) One has $\mathbf{P}(\Omega) = 1$.
(P_5) If (A_n) if a sequence of pairwise disjoint events, then $\mathbf{P}(\bigcup_{n \in \mathbf{N}} A_n) = \sum_{n \in \mathbf{N}} \mathbf{P}(A_n)$.
(P_6) If A is an event such that $\mathbf{P}(A) = 0$, then every subset B of A is an event.

Taking $A_n = \emptyset$ for every n, we see that $\mathbf{P}(\emptyset) = 0$.

Let A and B be disjoint events, and let us consider the sequence $(A, B, \emptyset, \emptyset, \ldots)$; we get $\mathbf{P}(A \cup B) = \mathbf{P}(A) + \mathbf{P}(B)$.

Taking $B = \Omega - A$, we also get $\mathbf{P}(\Omega - A) = 1 - \mathbf{P}(A)$.

Let A be an event, and let C be an event such that $C \subseteq A$.
Writing $A - C = (\Omega - C) \cap A$, we see that $A - C$ is an event.
On the other hand, we can write $\Omega - C = (\Omega - A) \cup (A - C)$; since the two events $\Omega - A$ and $A - C$ are disjoint, we have $1 - \mathbf{P}(C) = 1 - \mathbf{P}(A) + \mathbf{P}(A - C)$, hence

$$\mathbf{P}(A - C) = \mathbf{P}(A) - \mathbf{P}(C).$$

More generally, if A and B are events, then $A - (A \cap B)$ is an event which is disjoint from B and we have

$$\begin{aligned}\mathbf{P}(A \cup B) &= \mathbf{P}\big((A - (A \cap B)) \cup B\big) \\ &= \mathbf{P}(A - (A \cap B)) + \mathbf{P}(B) \\ &= \mathbf{P}(A) + \mathbf{P}(B) - \mathbf{P}(A \cap B).\end{aligned}$$

0.2.2. Discrete probabilities — Let Ω be a countable set (finite, for example).
Let us assume that the singleton $\{a\}$ is an event, for every $a \in \Omega$, and let us set $p_a = \mathbf{P}(\{a\})$. Then every subset A of Ω is an event, and we have

$$\mathbf{P}(A) = \sum_{a \in A} p_a.$$

In particular, $1 = \mathbf{P}(\Omega) = \sum_{a \in \Omega} p_a$.
Conversely, let $p: \Omega \to \mathbf{R}_+$ be a map such that

$$\sum_{a \in \Omega} p(a) = 1,$$

in the sense of summable families. For every subset A of Ω, let us define $\mathbf{P}(A) = \sum_{a \in A} p(a)$. One can check that $\mathfrak{P}(\Omega)$ is a σ-algebra and that $\mathbf{P}$ is a probability function on it.

Let us assume that the set Ω is finite. The equiprobable probability function on Ω is given by $p_a = 1/\operatorname{Card}(\Omega)$, for $a \in \Omega$. In this case, one has $\mathbf{P}(A) = \operatorname{Card}(A)/\operatorname{Card}(\Omega)$ for any event A: probability theory is a generalization of combinatorics.

Example 0.2.3 (Dice throws). — A classical example is given by throwing a 6-sided die, so that $\Omega = \{1, \ldots, 6\}$. If the die is fair, one has $p_a = 1/6$ for all $a \in \Omega$. The probability that the dice shows an even value is the probability of the event $\{2, 4, 6\}$, that is, $3/6 = 1/2$.

This example does not allow us to model a sequence of dice throws, but it can be generalized. If we wish, for example, to consider throws of five 6-sided dice (to study Yahtzee, for example), we shall set $\Omega = \{1, \ldots, 6\}^5$: this is the set of all 5-tuplets of elements of $\{1, \ldots, 6\}$. If the dice are fair and the throws are independent of one another, we will have $p(a_1, \ldots, a_5) = 1/6^5$ for any element $(a_1, \ldots, a_5) \in \Omega$.

As an exercise, we suggest computing the probability of getting a Yahtzee (five identical dice), a Four of a kind (four identical dice), a Full house (three identical dice, and the two remaining dice equal to another value), a Three of a kind (three identical dice), and a Straight (five consecutive dice).

If we have to study a large number of dice throws, still assumed to be fair and independent, we may consider $\Omega = \{1, \ldots, 6\}^{\mathbf{N}}$, the set of all infinite sequences $(a_0, a_1, \ldots,)$ from $\{1, \ldots, 6\}$. This is not a countable set, hence the necessity of the initial general framework, and one can show that we may endow it with a σ-algebra and a probability function such that, for every integer n and any finite sequence $(a_0, \ldots, a_{n-1})$, the set of all throws $(x_0, \ldots, x_{n-1})$ such that $x_m = a_m$ for $0 \leqslant m < n$ is an event of probability $1/6^n$.

0.3. Discrete random variables

Let **P** be a probability on a sample space Ω (we do not specify the σ-algebra of events).

Definition 0.3.1. — *A discrete random variable with values in a set* A *is a map* X: $\Omega \to$ A *which satisfies the following properties.*

a) *For every* $a \in$ A, *the set* $X^{-1}(a) = \{\omega \in \Omega\,;\, X(\omega) = a\}$ *is an event, denoted by* $(X = a)$.
b) *The family* $\big(\mathbf{P}(X = a)\big)_{a \in A}$ *is summable, with sum* 1.

In particular, the set of all $a \in$ A such that $\mathbf{P}(X = a) > 0$ is countable and nonempty. Its elements are called the *possible values* (more simply, the values) of the discrete random variable X. By definition, the set of all $\omega \in \Omega$ such that $X(\omega)$ is not a possible value of X is an event of zero probability.

The family $\big(\mathbf{P}(X = a)\big)_{a \in A}$ is the *law* of the discrete random variable X. Its support is the set of all possible values of the discrete random variable X. Conversely, a "discrete probability law" on the set A is a function p from A to $\mathbf{R}_+$ such that the family $(p(a))_{a \in A}$ is summable with sum 1.

When the set of all possible values of X is reduced to one element, say a, we say that X is *certain*, with value a.

Lemma 0.3.2. — a) *Let* X *and* Y *be discrete random variables on* Ω, *taking values in sets* A *and* B *respectively. Then the map* $\omega \mapsto \big(X(\omega), Y(\omega)\big)$ *is a discrete random variable on* Ω, *taking values in the product set* A × B.
b) *Let* X *be a discrete random variable on* Ω *with values in a set* A *and let* $f: A \to B$ *be a map. Then the map* $f \circ X$, *denoted by* $f(X)$, *is a discrete random variable on* Ω *with values in* B.

For example, the sum (or the product) of two discrete random variable with values in **C** is a discrete random variable. We apply assertion *b*) of the lemma to the discrete random variable (X, Y) and to the sum map $s: \mathbf{C}^2 \to \mathbf{C}$ (or to the product map $\pi: \mathbf{C}^2 \to \mathbf{C}$).

Proof. — a) Set $Z(\omega) = (X(\omega), Y(\omega))$. For $(a,b) \in A \times B$, the set $Z^{-1}(a,b)$ is equal to $(X = a) \cap (Y = b)$; it is therefore an event. For every finite subset C of $A \times B$, there exist finite subsets A_1 of A and B_1 of B such that $C \subseteq A_1 \times B_1$. Then

$$\sum_{(a,b) \in C} \mathbf{P}(X = a \text{ and } Y = b) \leq \sum_{a \in A_1} \sum_{b \in B_1} \mathbf{P}(X = a \text{ and } Y = b)$$

$$\leq \sum_{a \in A_1} \mathbf{P}(X = a) \leq 1.$$

This proves that the family $(\mathbf{P}(X = a \text{ and } Y = b))_{(a,b) \in A \times B}$ is summable. To prove that its sum is 1, we apply proposition 0.1.8 to this family and to the map $(a,b) \mapsto a$. Let $a \in A$. The set $(X = a)$ is the disjoint union of the countable family of events $(X = a \text{ and } Y = b)$, indexed by $b \in B$ which are possible values of Y, and of a set contained in the null event consisting of all $\omega \in \Omega$ such that $Y(\omega)$ is not a possible value of Y. Therefore,

$$\mathbf{P}(X = a) = \sum_{b \in B} \mathbf{P}(X = a \text{ and } Y = b).$$

By proposition 0.1.8, it follows that

$$\sum_{(a,b) \in A \times B} \mathbf{P}(X = a \text{ and } Y = b) = \sum_{a \in A} \mathbf{P}(X = a) = 1.$$

b) Let A' be the set of possible values of X, the set of all $a \in A$ such that $\mathbf{P}(X = a) > 0$, and let $B' = f(A')$. The set B' is a countable subset of B. For every $b \in B$, the set $f(X)^{-1}(b)$ is a disjoint union of the sets $X^{-1}(a)$, where a runs through $f^{-1}(b)$; it is the disjoint union of the countable family of all $X^{-1}(a)$, for $a \in A' \cap f^{-1}(b)$ and of a set contained in the null event $X^{-1}(A - A')$. Consequently, $f(X)^{-1}(b)$ is an event. This proves that $f(X)$ is a discrete random variable. □

0.3.3. Examples of laws — Let X be a discrete random variable.

a) One says that X has a *uniform law* if its set A of values is finite and if one has $\mathbf{P}(X = a) = 1/\mathrm{Card}(A)$ for every $a \in A$. One also says that X is a uniform random variable on A.

b) When the set of values of X is equal to the two-element set $\{0, 1\}$, the law of X is characterized by the real number $p = \mathbf{P}(X = 1) \in [0; 1]$; indeed, one then has $\mathbf{P}(X = 0) = 1 - p$. One then says that X follows a *Bernoulli law* with parameter p.

c) Let p be a real number in $[0; 1[$. A discrete random variable X follows a *geometric law* with parameter p if it takes its values in $\mathbf{N}^*$ and if $\mathbf{P}(X = n) = (1-p)p^{n-1}$ for every integer $n \geq 1$. (The limit case $p = 0$ corresponds to the case where X is certain with value 1.)

0.3. Discrete random variables.

d) Let p be a strictly positive real number. A discrete random variable W with values in **N** follows a *Poisson law* with parameter p if $\mathbf{P}(X = n) = e^{-p} p^n / n!$ for every integer n.

0.3.4. Expectation — Let X be a discrete random variable with values in **C**. One says that X admits an *expectation* if the family $(a\mathbf{P}(X = a))_{a \in \mathbf{C}}$ is summable; the expectation is then the sum, denoted by $\mathbf{E}(X)$, of this family.

If X admits an expectation, then the set of its possible values, that is, the set of all $a \in \mathbf{C}$ such that $\mathbf{P}(X = a) > 0$, is countable.

If X is certain, with value a, then X admits an expectation, equal to a.

We can generalize this definition when the set of possible values of X consists of positive real numbers: in this case, the expectation of X is defined as the sum, in $[0; +\infty]$, of the family $(a\mathbf{P}(X = a))_{a \geqslant 0}$ of positive real numbers; it is still denoted by $\mathbf{E}(X)$.

0.3.5. — Expectation satisfies the following properties.

a) *If X admits an expectation, then tX admits an expectation for every complex number t, and one has $\mathbf{E}(tX) = t\mathbf{E}(X)$.*

When $t = 0$, the random variable tX is certain, with value 0, and one has $\mathbf{E}(tX) = 0$. Otherwise, for every $a \in \mathbf{C}$, one has $\mathbf{P}(tX = a) = \mathbf{P}(X = a/t)$, so that $a\mathbf{P}(tX = a) = t(a/t)\mathbf{P}(X = a/t)$. The family $((a/t)\mathbf{P}(X = a/t))_{a \in \mathbf{C}}$ only differs from the family $(a\mathbf{P}(X = a))_{a \in \mathbf{C}}$ by the reindexing $a \mapsto a/t$; it is therefore summable, and has the same sum, that is, $\mathbf{E}(X)$. It follows that the family $(a\mathbf{P}(tX = a))_{a \in \mathbf{C}}$ is summable, and that its sum equals $t\mathbf{E}(X)$.

b) *If discrete random variables X and Y have an expectation, then the discrete random variable $X + Y$ admits an expectation, and one has $\mathbf{E}(X+Y) = \mathbf{E}(X) + \mathbf{E}(Y)$.*

Let A be the set of possible values of the random variable X, that is the set of $a \in \mathbf{C}$ such that $\mathbf{P}(X = a) > 0$; it is countable. Similarly, the set B of possible values of the random variable Y is countable. Let $s \colon \mathbf{C} \times \mathbf{C} \to \mathbf{C}$ be the sum map, given by $s(a, b) = a + b$, and let $C = s(A \times B) = A + B$; the set C is a countable subset of **C**.

For $c \in \mathbf{C}$, the set $(X+Y)^{-1}(c)$ is the disjoint union of events $X^{-1}(a) \cap Y^{-1}(b)$, where (a, b) runs among all pairs $(a, b) \in A \times B$ such that $a + b = c$, as well as of a subset of the null event $X^{-1}(\complement A) \cap Y^{-1}(\complement B)$. This implies that $(X+Y)^{-1}(c)$ is an event and that

$$\mathbf{P}(X+Y = c) = \sum_{\substack{(a,b) \in A \times B \\ a+b=c}} \mathbf{P}(X = a \text{ and } Y = b)$$

$$= \sum_{\substack{(a,b) \in C \times C \\ a+b=c}} \mathbf{P}(X = a \text{ and } Y = b).$$

Let us prove that the family $(a\,\mathbf{P}(X = a \text{ and } Y = b))_{(a,b) \in C \times C}$ is summable, with sum $\mathbf{E}(X)$. Indeed, let K be a finite subset of the product $\mathbf{C} \times \mathbf{C}$, and let K_1 be the set of $a \in A$ such that there exists a $b \in \mathbf{C}$ with $(a, b) \in K$. One has

$$\sum_{(a,b)\in K} |a\, P(X=a \text{ and } Y=b)| = \sum_{a\in K_1} |a| \sum_{\substack{b\in C \\ (a,b)\in K}} P(X=a \text{ and } Y=b)$$

$$\leq \sum_{a\in K_1} |a|\, P(X=a) \leq E(|X|),$$

which proves that the considered family is summable.

Let us apply proposition 0.1.8 to the map $(a,b) \mapsto a$ from $\mathbf{C}\times\mathbf{C}$ to $\mathbf{C}$; it comes

$$\sum_{(a,b)\in \mathbf{C}\times\mathbf{C}} a P(X=a \text{ and } Y=b) = \sum_{a\in \mathbf{C}} a \left(\sum_{b\in \mathbf{C}} P(X=a \text{ and } Y=b) \right).$$

For $a \in \mathbf{C}$, the event $(X = a)$ is the disjoint union of the countable family of events $(X = a \text{ and } Y = b)$, for $b \in B$, and of the null event $(X = a \text{ and } Y \notin B)$ (since it is contained in $(Y \notin B)$). This implies the equality

$$P(X=a) = \sum_{b\in B} P(X=a \text{ and } Y=b).$$

Therefore,

$$\sum_{(a,b)\in \mathbf{C}\times\mathbf{C}} a P(X=a \text{ and } Y=b) = \sum_{a\in \mathbf{C}} a\, P(X=a) = E(X).$$

Similarly, the family $(bP(X=a \text{ and } Y=b))_{(a,b)\in \mathbf{C}\times\mathbf{C}}$ is summable with sum $E(Y)$. This implies that the family $((a+b)P(X=a \text{ and } Y=b))_{(a,b)\in \mathbf{C}\times\mathbf{C}}$ is summable with sum $E(X) + E(Y)$.

Let us apply proposition 0.1.8 to the map s. For $c \in \mathbf{C}$, the family $((a+b)P(X=a \text{ and } Y=b))_{a+b=c}$ is summable, and its sum is equal to

$$\sum_{\substack{(a,b)\in \mathbf{C}\times\mathbf{C} \\ a+b=c}} (a+b)P(X=a \text{ and } Y=b) = c \sum_{\substack{(a,b)\in \mathbf{C}\times\mathbf{C} \\ a+b=c}} P(X=a \text{ and } Y=b)$$

$$= c\, P(X+Y=c).$$

Consequently, the family $(cP(X+Y=c))_{c\in \mathbf{C}}$ is summable and we have

$$E(X+Y) = \sum_{c\in \mathbf{C}} cP(X+Y=c)$$

$$= \sum_{c\in \mathbf{C}} \left(\sum_{\substack{(a,b)\in \mathbf{C}\times\mathbf{C} \\ a+b=c}} (a+b)P(X=a \text{ and } Y=b) \right)$$

$$= \sum_{(a,b)\in \mathbf{C}\times\mathbf{C}} (a+b)P(X=a \text{ and } Y=b)$$

$$= E(X) + E(Y).$$

0.3. Discrete random variables.

c) *If Y admits an expectation and if $|X| \leq Y$, then X and $|X|$ admit an expectation, and one has $|E(X)| \leq E(|X|) \leq E(Y)$.*

We first treat the case where the set A of possible values of X is finite and consists of positive real numbers. Then X has an expectation. Let us introduce the discrete random variable $Z = Y - X$. For $c \in \mathbf{C}$, the set $Z^{-1}(c)$ is the disjoint union of events $(X = a$ and $Y = c - a)$, for $a \in A$, and of a null event contained in $(X \notin A)$. Since $Y = X + Z$ and $Z \geq 0$, we get $E(Y) = E(X) + E(Z) \geq E(X)$.
Let us now assume that the possible values of X are positive real numbers, hence $0 \leq X \leq Y$. Let A be a finite subset of **C** and let $\varphi_A : \mathbf{C} \to \mathbf{C}$ be the function defined by $\varphi_A(a) = 0$ for $a \notin A$, and $\varphi_A(a) = a$ otherwise. Let X_A be the discrete random variable $\varphi_A(X)$; one has $X_A(\omega) = X(\omega)$ if $X(\omega) \in A$, and $X_A(\omega) = 0$ otherwise. Obviously, $0 \leq X_A \leq X \leq Y$, so that the first case applied to X_A gives $E(X_A) \leq E(Y)$. On the other hand, one has $P(X_A = a) = P(X = a)$ for $a \in A - \{0\}$, and $P(X_A = a) = 0$ for $a \notin A$. Consequently, $a\, P(X_A = a) = a\, P(X = a)$, for every $a \in A$, so that

$$\sum_{a \in A} a\, P(X = a) = E(X_A) \leq E(Y).$$

This implies that the family $(a\, P(X = a))_{a \in \mathbf{R}_+}$ is summable, and its sum is less than or equal to $E(Y)$. This proves that X admits an expectation, and that $E(X) \leq E(Y)$.
We finally prove the general case. By the preceding case, $|X|$ has an expectation, and $E(|X|) \leq E(Y)$. On the other hand, we may decompose X as a sum of four real positive discrete random variables,

$$X = \operatorname{Re}(X)^+ - \operatorname{Re}(X)^- + i \operatorname{Im}(X)^+ - i \operatorname{Im}(X)^-,$$

and each of them is less or equal than Y. By what precedes, they all have an expectation, hence X admits an expectation.
It remains to prove the inequality $|E(X)| \leq E(|X|)$. With that aim, let us apply proposition 0.1.8 to the family $(a\, P(X = a))$ and to the map $|\cdot|$ from **C** to **R**. For every $r \in \mathbf{R}_+$, the event $(|X| = r)$ is the disjoint union of a countable family of events $(X = a)$, where a runs among all possible values a of X such that $|a| = r$, and of a null event. We then have

$$P(|X| = r) = \sum_{\substack{a \in A \\ |a| = r}} P(X = a).$$

Then

$$E(X) = \sum_{a \in \mathbf{C}} a\, P(X = a) = \sum_{r \in \mathbf{R}_+} \sum_{|a| = r} a\, P(X = a).$$

For $r \in \mathbf{R}_+$, we also have

$$\Big| \sum_{|a|=r} a\mathbf{P}(X=a) \Big| \leq \sum_{|a|=r} |a|\mathbf{P}(X=a)$$
$$= r \sum_{|a|=r} \mathbf{P}(X=a) = r\,\mathbf{P}(|X|=r).$$

Subsequently,
$$|\mathbf{E}(X)| \leq \sum_{r \in \mathbf{R}_+} r\mathbf{P}(|X|=r) = \mathbf{E}(|X|),$$
as was to be proved.

0.3.6. Moments, variance — Let k be a natural integer. One says that a discrete random variable has a *moment* of order k if the random variable X^k admits an expectation, or, equivalently, if the expectation of the real positive valued random variable $|X|^k$ is finite; this expectation is then called the *moment of order k* of the discrete random variable X.

By definition, having a moment of order 1 is the same as having an expectation.

When the discrete random variable X admits moments of order 1 and 2, one defines its *variance* by
$$\mathbf{V}(X) = \mathbf{E}\big(|X - \mathbf{E}(X)|^2\big). \tag{0.3.6.1}$$
Since $|X - \mathbf{E}(X)|^2 = |X|^2 - \overline{X}\mathbf{E}(X) - X\overline{\mathbf{E}(X)} + |\mathbf{E}(X)|^2$, we also have
$$\mathbf{V}(X) = \mathbf{E}(|X|^2) - \mathbf{E}(\overline{X})\mathbf{E}(X) - \mathbf{E}(X)\overline{\mathbf{E}(X)} + |\mathbf{E}(X)|^2$$
$$= \mathbf{E}(|X|^2) - |\mathbf{E}(X)|^2. \tag{0.3.6.2}$$

Proposition 0.3.7. — *Let X and Y be discrete random variables with values in $\mathbf{C}$ which admit a moment of order 2.*

a) (Young) The random variable XY has an expectation, and one has
$$|\mathbf{E}(XY)| \leq \frac{1}{2}\big(\mathbf{E}(|X|^2) + \mathbf{E}(|Y|^2)\big). \tag{0.3.7.1}$$

b) (Cauchy–Schwarz) The random variable XY has an expectation, and one has
$$|\mathbf{E}(XY)| \leq \mathbf{E}(|X|^2)^{1/2}\,\mathbf{E}(|Y|^2)^{1/2}. \tag{0.3.7.2}$$

c) In particular, the random variable X has an expectation, and one has $|\mathbf{E}(X)| \leq \mathbf{E}(|X|^2)^{1/2}$.

d) (Minkowski) One has
$$\mathbf{E}(|X+Y|^2)^{1/2} \leq \mathbf{E}(|X|^2)^{1/2} + \mathbf{E}(|Y|^2)^{1/2}. \tag{0.3.7.3}$$

Proof. — a) Up to replacing X and Y by $|X|$ and $|Y|$, we may assume that they take positive real values. For real numbers x and y, one has $x^2 + y^2 - 2xy = (x-y)^2 \geq 0$, so that $xy \leq (x^2+y^2)/2$. Consequently, for any $\omega \in \Omega$, one has

0.4. Independence, conditional expectation.

$$|XY(\omega)| = |X(\omega)| \times |Y(\omega)| \leq \frac{1}{2}|X(\omega)|^2 + \frac{1}{2}|Y(\omega)|^2,$$

hence the inequality of positive real valued random variables

$$|XY| \leq \frac{1}{2}(|X|^2 + |Y|^2).$$

Since X^2 and Y^2 admits an expectation, so does XY, and

$$\mathbf{E}(|XY|) \leq \frac{1}{2}\left(\mathbf{E}(|X|^2) + \mathbf{E}(|Y|^2)\right).$$

b) Let a and b be strictly positive real numbers such that $\mathbf{E}(|X|^2) \leq a^2$ and $\mathbf{E}(|Y|^2) \leq b^2$. Let us apply Young's inequality to the random variables X/a and Y/b. We get

$$\frac{1}{ab}\mathbf{E}(|XY|) \leq \frac{1}{2a^2}\mathbf{E}(|X^2|) + \frac{1}{2b^2}\mathbf{E}(|Y|^2) \leq 1,$$

hence $\mathbf{E}(|XY|) \leq ab$. When a tends to $\mathbf{E}(|X|^2)^{1/2}$ and b to $\mathbf{E}(|Y|^2)^{1/2}$, we obtain the inequality

$$\mathbf{E}(|XY|) \leq \mathbf{E}(|X|^2)^{1/2}\mathbf{E}(|Y|^2)^{1/2}.$$

c) Let Y be the certain random variable with value 1; since $\mathbf{E}(1) = \mathbf{E}(1^2) = 1$, it admits a moment of order 2, and one has $\mathbf{E}(|X|) = \mathbf{E}(|XY|) \leq \mathbf{E}(|X|^2)^{1/2}$.

d) One has $|X + Y|^2 \leq |X|^2 + 2|X|\,|Y| + |Y|^2$, so that

$$\mathbf{E}(|X+Y|^2) \leq \mathbf{E}(|X|^2) + 2\mathbf{E}(|X|\,|Y|) + \mathbf{E}(|Y|^2)$$
$$\leq \mathbf{E}(|X|^2) + 2\mathbf{E}(|X|^2)^{1/2}\mathbf{E}(|Y|^2)^{1/2} + \mathbf{E}(|Y|^2)$$
$$= \left(\mathbf{E}(|X|^2)^{1/2} + \mathbf{E}(|Y|^2)^{1/2}\right)^2,$$

hence Minkowski inequality. □

0.4. Independence, conditional expectation

0.4.1. — Let **P** be a probability on a sample space Ω and let A, B be events such that $\mathbf{P}(B) > 0$. One defines the *conditional probability* of A given B by

$$\mathbf{P}(A \mid B) = \frac{\mathbf{P}(A \cap B)}{\mathbf{P}(B)}.$$

Observe that $A \mapsto \mathbf{P}(A \mid B)$ is a probability $\mathbf{P}_B$ on the sample space Ω (for the same σ-algebra as **P**).

One says that events A and B are *independent* if $\mathbf{P}(A \cap B) = \mathbf{P}(A)\mathbf{P}(B)$. More generally, one says that a family (A_i) of events, finite or infinite, is independent, if for every finite subset I of indices, one has

$$\mathbf{P}\Big(\bigcap_{i \in I} A_i\Big) = \prod_{i \in I} \mathbf{P}(A_i).$$

Proposition 0.4.2 (Bayes's law). — *Let A and B be events such that $\mathbf{P}(A) > 0$ and $\mathbf{P}(B) > 0$. One has*

$$\mathbf{P}(A \mid B) = \frac{\mathbf{P}(B \mid A)\,\mathbf{P}(A)}{\mathbf{P}(B)}.$$

Proof. — By definition, one has $\mathbf{P}(A \mid B) = \mathbf{P}(A \cap B)/\mathbf{P}(B)$ and $\mathbf{P}(B \mid A) = \mathbf{P}(A \cap B)/\mathbf{P}(A)$. Consequently, $\mathbf{P}(B \mid A)\,\mathbf{P}(A) = \mathbf{P}(A \cap B)$, hence

$$\frac{\mathbf{P}(B \mid A)\,\mathbf{P}(A)}{\mathbf{P}(B)} = \frac{\mathbf{P}(A \cap B)}{\mathbf{P}(B)} = \mathbf{P}(A \mid B).$$

This proves the formula. □

Proposition 0.4.3 (Law of total probabilities). — *Let (B_n) be a sequence (finite or infinite) of pairwise disjoint events such that $\Omega = \bigcup B_n$ and $\mathbf{P}(B_n) > 0$ for all n. For every event A, one has*

$$\mathbf{P}(A) = \sum_n \mathbf{P}(A \mid B_n)\mathbf{P}(B_n).$$

Proof. — By definition of conditional probabilities, one has $\mathbf{P}(A \mid B_n)\mathbf{P}(B_n) = \mathbf{P}(A \cap B_n)$. The events $A \cap B_n$ are pairwise disjoint and their union is A, since $\Omega = \bigcup B_n$. It follows that $\mathbf{P}(A) = \sum_n \mathbf{P}(A \cap B_n)$. □

Definition 0.4.4. — *Let X and Y be discrete random variables. One says that X and Y are* independent *if for every a, b, one has*

$$\mathbf{P}(X = a \text{ and } Y = b) = \mathbf{P}(X = a)\,\mathbf{P}(Y = b).$$

Proposition 0.4.5. — *Let X and Y be discrete random variables with complex values; assume that they are independent. If X and Y admit an expectation, then their product XY admits an expectation, and one has $\mathbf{E}(XY) = \mathbf{E}(X)\mathbf{E}(Y)$.*

Proof. — Let C be a finite subset of $\mathbf{C}^2$ and let A, B be finite subsets of $\mathbf{C}$ such that $C \subseteq A \times B$. Then

$$\sum_{(a,b) \in C} |ab\mathbf{P}(X = a \text{ and } Y = b)| \leqslant \sum_{a \in A} \sum_{b \in B} |a||b|\mathbf{P}(X = a)\mathbf{P}(Y = b)$$

$$\leqslant \mathbf{E}(|X|)\,\mathbf{E}(|Y|).$$

This proves that the family $\big(ab\mathbf{P}(X = a \text{ and } Y = b)\big)_{(a,b) \in \mathbf{C}^2}$ is summable.

0.4. Independence, conditional expectation. 19

Let us then apply proposition 0.1.8 to this family and to the map $(a, b) \mapsto a$ from $\mathbf{C}^2$ to $\mathbf{C}$. Let $a \in \mathbf{C}$. One has

$$\sum_{b \in \mathbf{C}} ab\mathbf{P}(X = a \text{ and } Y = b) = a\mathbf{P}(X = a) \sum_{b \in \mathbf{C}} b\mathbf{P}(Y = b) = a\mathbf{P}(X = a)\mathbf{E}(Y).$$

Therefore,

$$\sum_{(a,b) \in \mathbf{C}^2} ab\mathbf{P}(X = a \text{ and } Y = b) = \sum_{a \in \mathbf{C}} a\mathbf{P}(X = a)\mathbf{E}(Y) = \mathbf{E}(X)\mathbf{E}(Y).$$

Let us now apply proposition 0.1.8 to the same family, but to the map $(a, b) \mapsto ab$. Let A be the set of possible values of X and let B be the set of possible values of Y. Let $c \in \mathbf{C}$. The event $(XY = c)$ is the disjoint union of the countable family of events $(X = a \text{ and } Y = b)$, where (a, b) runs among all pairs $(a, b) \in A \times B$, and of a null event contained in $(X \notin A) \cup (Y \notin B)$. It follows that

$$\sum_{\substack{(a,b) \in \mathbf{C} \times \mathbf{C} \\ ab=c}} ab\mathbf{P}(X = a \text{ and } Y = b) = c \sum_{\substack{(a,b) \in \mathbf{C} \times \mathbf{C} \\ ab=c}} \mathbf{P}(X = a)\mathbf{P}(Y = b)$$

$$= c\,\mathbf{P}(XY = c),$$

so that

$$\mathbf{E}(X)\mathbf{E}(Y) = \sum_{(a,b) \in \mathbf{C}^2} ab\,\mathbf{P}(X = a \text{ and } Y = b)$$

$$= \sum_{c \in \mathbf{C}} c\,\mathbf{P}(XY = c) = \mathbf{E}(XY). \square$$

0.4.6. — Let X be a discrete random variable that admits an expectation and let B be an event such that $\mathbf{P}(B) > 0$. For every $a \in \mathbf{C}$, one has

$$\mathbf{P}(X = a \mid B) = \frac{\mathbf{P}\big((X = a) \cap B\big)}{\mathbf{P}(B)} \leqslant \frac{\mathbf{P}(X = a)}{\mathbf{P}(B)}.$$

Consequently, the family $\big(a\mathbf{P}(X = a \mid B)\big)_{a \in \mathbf{C}}$ is summable. Its sum is called the *conditional expectation of* X *given* B, and is denoted by $\mathbf{E}(X \mid B)$.

We may observe that it is the expectation of the random variable X for the probability $\mathbf{P}_B = \mathbf{P}(\cdot \mid B)$ on Ω. In particular, it satisfies the following properties.

1. For every $t \in \mathbf{C}$, one has $\mathbf{E}(tX \mid B) = t\mathbf{E}(X \mid B)$.
2. If Y is a discrete random variable that admits a conditional expectation given B, then so does $X + Y$, and one has

$$\mathbf{E}(X + Y \mid B) = \mathbf{E}(X \mid B) + \mathbf{E}(Y \mid B).$$

3. If $|X| \leqslant Y$, then $|\mathbf{E}(X \mid B)| \leqslant \mathbf{E}(|X| \mid B) \leqslant \mathbf{E}(Y \mid B)$.

0.4.7. — More generally, let us say that a family (X_i) of discrete random variables is independent or, by abuse of language, that the random variables X_i are independent, if for every finite subset I of indices and every $(a_i)_{i \in I}$, one has

$$\mathbf{P}\left(\bigcap_{i \in I}(X_i = a_i)\right) = \prod_{i \in I} \mathbf{P}(X_i = a_i).$$

If the family is indexed by integers ≥ 1, this amounts to saying that for every n, the random variable X_n is independent of the discrete random variable $(X_1, \ldots, X_{n-1})$.

Let us consider a finite sequence $(X_1, \ldots, X_n)$ of discrete random variables with complex values, which is independent. If they have an expectation, then so does their product $X_1 \cdots X_n$ and

$$\mathbf{E}(X_1 \cdots X_n) = \mathbf{E}(X_1) \cdots \mathbf{E}(X_n).$$

This follows by induction from the case of two random variables.

0.4.8. — Let X and Y be discrete random variables. Let us assume that X takes complex values and admits an expectation. One then defines as follows a discrete random variable $\mathbf{E}(X \mid Y)$ on Ω.

Let $\omega \in \Omega$. If $\mathbf{P}(Y = Y(\omega)) > 0$, that is, if $Y(\omega)$ is a possible value of Y, we set $\mathbf{E}(X \mid Y)(\omega) = \mathbf{E}(X \mid Y = Y(\omega))$; otherwise, we set $\mathbf{E}(X \mid Y)(\omega) = 0$.

One can prove that it is indeed a discrete random variable. It is called the *conditional expectancy of* X *given* Y.

In particular, for every $a \in \mathbf{C}$, the set of all $\omega \in \Omega$ such that $\mathbf{E}(X \mid Y)(\omega) = a$ is an event.

One defines in an analogous way the *conditional variance* of X given Y,

$$\mathbf{V}(X \mid Y)(\omega) = \begin{cases} \mathbf{V}(X \mid Y = Y(\omega)), & \text{if } \mathbf{P}((Y = Y(\omega))) > 0, \\ 0, & \text{otherwise.} \end{cases}$$

Again, this is a discrete random variable on Ω.

Exercises

Exercise 0.1 (Uniform law). [p. 139] Let $n \in \mathbf{N}^*$, and let X be a uniform random variable on $\{1, \ldots, n\}$, meaning that all probabilities $\mathbf{P}(X = a)$ are equal to $1/n$, for $a \in \{1, \ldots, n\}$.

   a) Recall the values of $\mathbf{P}(X = a)$ for $a \in \{1, \ldots, n\}$.

   b) Compute the probability that X is even. More generally, if d is an integer ≥ 1, compute the probability that X is a multiple of d.

   c) Compute the expectation of X.

   d) Compute the variance of X.

0.4. Independence, conditional expectation.

Exercise 0.2. [p. 140]   *a)* Three prisoners are at risk of execution.. The warder tells one of them that one of them has been pardoned, but refuses to tell him which; however he gives him the name of one of the two prisoners that shall be executed (but not his own). Should the prisoner feel reassured by hearing the name of one of his inmates?

b) On the stage of a television game show are three doors, behind one of which is a car (the car is supposed to be a valued prize, the TV networks don't care about global warming or the energy crisis). The TV host asks the candidate to choose one of the three doors he would like to open, but then (randomly) opens one of the two other doors that does not hide the car. If the candidate wishes to win the car (pretend our candidate doesn't care either...) should he open the door he initially chose, or should he open the third door?

Exercise 0.3 (Poisson law). [p. 141]   Let p be a positive real number. Let X be a discrete random variable which follows the Poisson law of parameter p: this means that it is integer-valued and that for any integer $n \geq 0$, one has $\mathbf{P}(X = n) = e^{-p} p^n / n!$.

a) Compute the expectation of X.

b) Compute the expectation of X^2 and the variance of X.

c) Prove that there exists, for every integer $k \geq 0$, a polynomial L_k of degree k such that the expectation of $\mathbf{E}(X^k) = L_k(p)$.

Exercise 0.4 (Geometric law). [p. 143]   Let p and q be strictly positive real numbers such that $p + q = 1$. Let X be a discrete random variable whose law is the geometric law of parameter p, that is, $\mathbf{P}(X = n) = q\, p^{n-1}$ for every integer $n \geq 1$.

a) Compute the expectation of X.

b) Compute the variance of X.

c) Prove that there exist, for every integer $k \geq 0$, a monic polynomial of degree k such that $\mathbf{E}(X^k) = S_k(p)/q^k$.

Exercise 0.5. [p. 144]
A cheat has an unfair 6-sided die whose result is 1 with probability 2/3 and any other face with probability 1/15. However, his die is in a bag that contains two fair 6-sided dice for which each face appears with probability 1/6. These three dice look identical.

a) Our cheat takes one of the three dice at random. What is the probability that he picked up the unfair one? (One may introduce the random variable A, with values true/false, saying that "the cheat has chosen the unfair die".)

b) The cheat throws the die, and gets 1. Conditionally to this result, what is the probability that he took the unfair die? (Introduce the random variable X which gives the value of the die.)

c) He throws this die again, and gets 1 again. What now is the probability that he took the unfair die? (Introduce the random variable Y that gives the value of the second throw, and justify that X and Y are independent conditionally to A.)

Exercise 0.6. [p. 145]   For the first three questions, we assume that X and Y are the outcomes of two independent throws of a fair 6-sided die, and we set $Z = X + Y$.
  a) Compute the law of Z.
  b) For every $a \in \{1, \ldots, 12\}$, compute $\mathbf{E}(X \mid Z = a)$.
  c) Conclude that $\mathbf{E}(X \mid Z) = Z/2$.
  d) More generally, let X and Y be discrete random variables with complex values, and let $Z = X + Y$. Assuming that X and Y are independent and follow the same law, prove that $\mathbf{E}(X \mid Z) = Z/2$.
  e) Assume that X and Y are independent and follow a Bernoulli law of parameter o. Compute $\mathbf{E}(X \mid XY = a)$ for all a.

Exercise 0.7 (Conditional expectation, conditional variance). [p. 147]   Let X and Y be discrete random variables with values in **C**.
  a) Assume that X admits an expectation. Prove that the random variable $\mathbf{E}(X \mid Y)$ admits an expectation and that $\mathbf{E}\big(\mathbf{E}(X \mid Y)\big) = \mathbf{E}(X)$.
  b) Assume that X admits a variance. Prove that $\mathbf{V}(X) = \mathbf{E}\big(\mathbf{V}(X \mid Y)\big) + \mathbf{V}\big(\mathbf{E}(X \mid Y)\big)$.

Chapter 1.
Entropy and mutual information

Entropy is a fundamental concept in information theory, proposed by SHANNON (1948). We shall first define the entropy of a discrete random variable, and show on some examples that it can be interpreted in two ways, either as an amount of randomness, or as an amount of information. This duality information/randomness may look mysterious, but it is at the heart of the probabilistic formalization of phenomena for which some unknown information is mathematically treated as if it were random. That this point of view is at all pertinent is not a priori obvious, but is attested by the importance of probability theory in the mathematics of the second half of the 20th century and by its impact in applications, the last one being probably the fantastic (and frightening) development of supervised learning. Two variants of entropy appear as being very important, both for their efficiency in computations and for their interpretation in information theory: *conditional entropy* and *mutual information*.

In information theory, a signal is often assumed to be a sequence of symbols, or elementary signals, emitted at regularly spaced times, and which probability theory represents as a sequence of random variables—a *stochastic process*. This gives rise to the notion of *entropy rate:* not the global entropy of the signal, but its entropy by time unit, or by symbol.

Among the simplest stochastic processes, one can find those which are totally independent. Closer to real phenomena that we may want to model are the *Markov processes*: these are those for which the symbol emitted at time $n+1$ only depends on that emitted at time n. Of course, it is an independence condition in the sense of probability theory. One can analyse the entropy rate of these processes, in particular when they are stationary or ergodic.

1.1. Entropy of a discrete random variable

1.1.1. — We denote here by log : $\mathbf{R}_{>0} \to \mathbf{R}$ the usual, "natural", logarithm function, inverse to the exponential function, in other words, the Napierian logarithm. It is a $\mathscr{C}^\infty$ function; since its derivative is the strictly positive function $x \mapsto 1/x$, the logarithm is strictly increasing; it obeys the functional equation

$$\log(xy) = \log(x) + \log(y), \qquad \text{for all } x, y > 0. \tag{1.1.1.1}$$

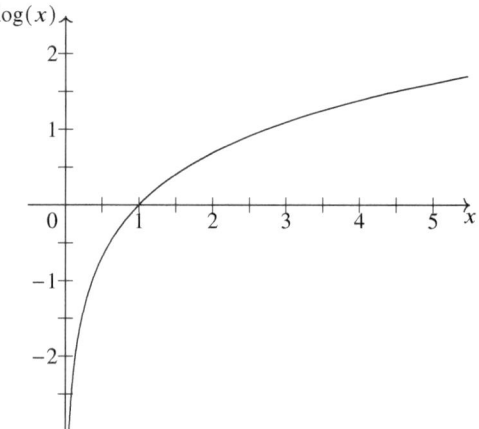

Fig. 1.1 Graph of the natural logarithm

We know the limits

$$\lim_{x \to 0^+} \log(x) = -\infty, \qquad \lim_{x \to +\infty} \log(x) = +\infty, \tag{1.1.1.2}$$

as well as the limits, for every real number $\alpha > 0$:

$$\lim_{x \to 0^+} x^\alpha \log(x) = 0, \qquad \lim_{x \to +\infty} x^{-\alpha} \log(x) = 0. \tag{1.1.1.3}$$

Its second derivative, the function $x \mapsto -1/x^2$, is strictly negative. Consequently, the logarithm function is strictly *concave*: its graph is below its tangents, and above its chords.

1.1.2. — If a is a real number > 0, the "logarithm base a" is the function given by

$$\log_a(x) = \frac{\log(x)}{\log(a)}.$$

1.1. Entropy of a discrete random variable.

It satisfies similar properties to the natural logarithm, which corresponds to the case where $a = e = 2.718\,281\,828\ldots$ is Euler's number.[1] The case $a = 10$ is very customary in physics; in information theory, it will be useful to take $a = 2$.

1.1.3. — The function $x \mapsto -x\log(x)$ from $]0;1]$ to $\mathbf{R}$ takes positive real values. It has limit 0 at 0, so that we extend it by continuity at 0, assigning it the value 0. This also justifies the writing convention $0 \times \log(0) = 0$, itself the convention $0^0 = 1$ in disguise! This function is also infinitely differentiable on $]0;1[$, with derivative $x \mapsto -\log(x) - 1 = -\log(ex)$; it is therefore strictly increasing on the interval $[0;1/e]$ and strictly decreasing on the interval $[1/e;1]$. Since its second derivative is the function $x \mapsto -1/x$, which takes strictly negative values on $]0;1]$, this function is strictly concave.

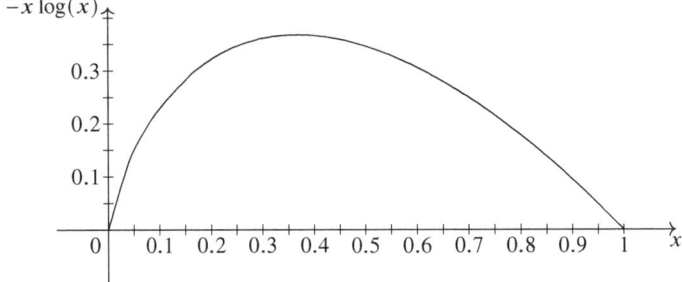

Fig. 1.2 Graph of the function $x \mapsto -x\log(x)$

Definition 1.1.4. — The *entropy of a discrete random variable* X *is defined by*

$$H(X) = \sum_x \Big(-\mathbf{P}(X = x) \log \big(\mathbf{P}(X = x)\big)\Big). \qquad (1.1.4.1)$$

The entropy is thus defined as the sum of a family of real numbers, indexed by the set of all values x of the random variable X. In this definition, we use the convention $0 \log(0) = 0$; we may thus only consider, if we prefer so, the values x for which $\mathbf{P}(X = x) > 0$, that is, the possible values of X. We have also seen in section 1.1.3 that this family has positive terms. Its sum is then well defined as an element of $[0; +\infty]$. It is finite if X only has finitely many possible values.

We may of course define entropy in any base $a > 0$.

$$H_a(X) = \sum_x \Big(-\mathbf{P}(X = x) \log_a \big(\mathbf{P}(X = x)\big)\Big) = \frac{H(X)}{\log(a)}. \qquad (1.1.4.2)$$

---

[1] This number has been discovered by Jacob BERNOULLI in 1683, related to a problem of compound interest, although it was implicitly present in a table of (natural) logarithms published by John NAPIER in 1618. The notation itself is due to Leonhard EULER who started to use it around 1730.

Example 1.1.5 (Dice throws). — Let us consider a fair 6-sided die. The probability that each face shows up is thus $1/6$; the entropy of the corresponding random variable is then equal to $6 \times \left(-\frac{1}{6} \log(\frac{1}{6})\right) = \log(6)$.

More generally, a uniform random variable on an N-element set has entropy $\log(N)$.

Let us consider that N is a power of 2, say $N = 2^n$, and the possible values of X are the integers $\{0, \ldots, N-1\}$. Then one can know the outcome of X by asking successively n "binary questions", namely the digits of the base 2 expansion of X. In this case, one has $H_2(X) = \log_2(N) = n$.

We shall see how Shannon's theorems interpret (up to a unit) the base 2 entropy of a random variable as the average number of binary questions that one has to ask to know its outcome.

Let us now consider two fair 6-sided dice and let Y be the random variable corresponding to the sum of the two faces that show up. It may take the values $2, 3, \ldots, 12$; the value 2 is only possible for the throws $(1, 1)$, the value 3 corresponds to the two throws $(1, 2)$ and $(2, 1)$, etc. Consequently, the probabilities of the events $(X = x)$ are subsumed by the following table.

x	2	3	4	5	6	7	8	9	10	11	12
$P(X=x)$	1/36	2/36	3/36	4/36	5/36	6/36	5/36	4/36	3/36	2/36	1/36

Fig. 1.3 Probabilities for the sum of two dice

Its base 2 entropy is then equal to

$$H_2(Y) = -\frac{1}{36} \log_2\left(\frac{1}{36}\right) - \cdots - \frac{1}{36} \log_2\left(\frac{1}{36}\right) \approx 3.274$$

while the entropy of a uniform random variable in $\{2, \ldots, 12\}$ is equal to

$$\log_2(11) \approx 3.459.$$

There is a bit less randomness in the outcome of the sum of two dice than in the outcome of a fair "11-sided" die indicating the integers from 2 to 12.

Example 1.1.6 (Bernoulli random variables). — Let p be a real number in $[0; 1]$. Recall that a random variable X is said to follow the *Bernoulli law*[2] of parameter p if it takes the value 1 with probability p and the value 0 with probability $1 - p$. The entropy of such a random variable is thus given by

---

[2] Jacob (1655–1705) and Johann (1667–1748) BERNOULLI were two Swiss scientists whose works encompassed infinitesimal calculus, the calculus of variations, probability theory, combinatorics... A third member of the family, Daniel (1700–1782), was also interested in fluid mechanics and economics.

1.1. Entropy of a discrete random variable.

$$h(p) = \begin{cases} -p\log(p) - (1-p)\log(1-p) & \text{if } 0 < p < 1, \\ 0 & \text{if } p = 0 \text{ or } p = 1. \end{cases}$$

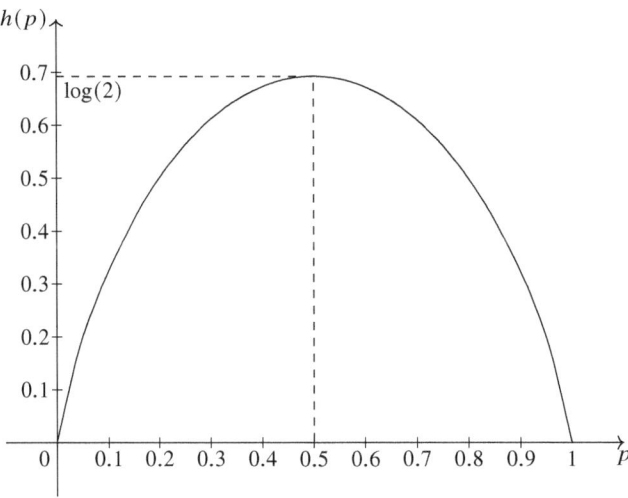

Fig. 1.4 Graph of the entropy of a Bernoulli random variable

Since the function $x \mapsto -x\log(x)$ is continuous on $[0;1]$, infinitely differentiable on $]0;1[$, with derivative $x \mapsto -\log(x) - 1$, the function h is continuous on $[0;1]$, infinitely differentiable on $]0;1[$, with derivative

$$h'(p) = -\log(p) + \log(1-p).$$

Its second derivative is given by

$$h''(p) = -\frac{1}{p} - \frac{1}{1-p}$$

hence is strictly negative on $]0;1[$, so that the function h is strictly concave.

One has $h'(1/2) = 0$, so that $h'(p) > 0$ for $p \in]0;1/2[$ and $h'(p) < 0$ for $p \in]1/2;1[$. The function h is strictly increasing on $[0;1/2]$ and strictly decreasing on $[1/2;1]$. It reaches its maximum at $p = 1/2$, with value $h(1/2) = \log(2)$.

One can see here the interest of base 2 entropies: the function h_2 defined by

$$h_2(p) = h(p)/\log(2) = -p\log_2(p) - (1-p)\log_2(1-p)$$

has for its image the interval $[0;1]$; its maximum is 1.

1.2. Conditional entropy

1.2.1. — Let A be an event of nonzero probability. Setting

$$\mathbf{P}_A(B) = \mathbf{P}(B \mid A) = \frac{\mathbf{P}(B)}{\mathbf{P}(A)}$$

for every event B which is contained in A, we may view A as a sample space with probability function $\mathbf{P}_A$. If X is a discrete random variable, we may then *condition* it to A by considering its restriction to A, here denoted by X | A.

1.2. Conditional entropy.

Definition 1.2.2. — *Let X and Y be discrete random variables. The expression*

$$H(X \mid Y) = \sum_{y} P(Y = y) H(X \mid Y = y) \tag{1.2.2.1}$$

is called the conditional entropy *of X given Y (or relative to Y).*

In principle, the random variable $X \mid \{Y = y\}$ is only defined when $P(Y = y) \neq 0$; if $P(Y = y) = 0$, we just remove the corresponding term from the sum. As the sum of a family of (possibly infinite) positive real numbers, $H(X \mid Y)$ is an element of $[0; +\infty]$. This sum is infinite if there exists a y such that $P(Y = y) > 0$ and $H(X \mid Y = y) = +\infty$; if no such y exists, it is still possible that the sum is $+\infty$. In any case, if X and Y only have finitely many possible values, the conditional entropy of X given Y will be finite.

If the random variable Y is certain, that is, if there exists a y such that $P(Y = y) = 1$, then $H(X \mid Y) = H(X \mid Y = y) = H(X)$.

Proposition 1.2.3. — *Let X and Y be discrete random variables. We have*

$$H(X, Y) = H(Y) + H(X \mid Y). \tag{1.2.3.1}$$

This is a kind of chain rule for entropy which, in the French tradition, is often referred to as "Chasles' rule"[3] for entropy.

Proof. — Let us start from the definitions of entropy and of conditional entropy. To simplify our notation, we write:

$$p_x = P(X = x), \quad q_y = P(Y = y), \quad r_{xy} = P(X = x \text{ and } Y = y).$$

Summing over all possible values of Y, that is all y such that $q_y > 0$, we then have:

$$H(Y) + H(X \mid Y) = \sum_{y} -q_y \log(q_y) + \sum_{y} q_y H(X \mid Y = y).$$

On the other hand, for every y such that $q_y > 0$, the law of the random variable $X \mid \{Y = y\}$ is given by

$$P(X = x \mid Y = y) = \frac{P(X = x \text{ and } Y = y)}{P(Y = y)} = \frac{r_{xy}}{q_y},$$

hence its entropy is equal to

---

[3] Michel CHASLES (1793–1880) was a specialist in geometry. We owe him to have promoted the use of the cross-ratio ("anharmonic ratio" in his terminology) and the solution of Steiner's "conic problem" consisting in counting the number of plane conics which are tangent to five given conics (in general, there are 3264 of them!). In a way that does not do justice to his mathematical talent, his name is often associated with additivity relations such as $\int_a^b f + \int_b^c f = \int_a^c f$. Chasles is also known for having been the naive victim of forger Denis Vrain-Lucas, who sold him many fake letters of celebrities, including one from Cleopatra to Julius Caesar!

$$H(X \mid Y = y) = \sum_x -P(X = x \mid Y = y) \log \left(P(X = x \mid Y = y)\right)$$

$$= \sum_x -\frac{r_{xy}}{q_y} \log \left(\frac{r_{xy}}{q_y}\right).$$

Summing termwise these two families with positive real terms, we get

$$H(Y) + H(X \mid Y) = \sum_y q_y \log \left(\frac{1}{q_y}\right) + \sum_{x,y} r_{xy} \log \left(\frac{q_y}{r_{xy}}\right).$$

Since $q_y = P(Y = y) = \sum_x P(X = x$ and $Y = y) = \sum_x r_{xy}$, we get

$$H(Y) + H(X \mid Y) = \sum_{x,y} \left(r_{xy} \log \left(\frac{1}{q_y}\right) + r_{xy} \log \left(\frac{q_y}{r_{xy}}\right)\right)$$

$$= \sum_{x,y} r_{xy} \log \left(\frac{1}{r_{xy}}\right) = H(X, Y),$$

by definition of the entropy of the pair (X, Y). □

Corollary 1.2.4. — *Let* X, Y, Z *be discrete random variables. One has*

$$H(X, Y \mid Z) = H(Y \mid Z) + H(X \mid Y, Z).$$

Proof. — Let z be a possible value of Z. Let us apply the preceding proposition to the random variables $X \mid (Z = z)$ and $Y \mid (Z = z)$; we get

$$H(X, Y \mid Z = z) = H(Y \mid Z = z) + H(X \mid Y, Z = z).$$

Let us multiply this relation by $P(Z = z)$ and compute their sum, for all possible values z. By definition of the conditional entropies $H(X, Y \mid Z)$ and $H(Y \mid Z)$, we obtain

$$H(X, Y \mid Z) = H(Y \mid Z) + \sum_z P(Z = z) H(X \mid Y, Z = z).$$

To compute this last term, let us go back to the definition of the conditional entropy $H(X \mid Y, Z = z)$; for each z, we have

$$H(X \mid Y, Z = z) = \sum_y P(Y = y \mid Z = z) H(X \mid Y = y, Z = z),$$

so that

$$P(Z = z) H(X \mid Y, Z = z)$$

$$= \sum_y \sum_{\substack{z \\ P(Z=z)>0}} P(Z = z) P(Y = y \mid Z = z) H(X \mid Y = y, Z = z)$$

1.2. Conditional entropy.

$$= \sum_y \sum_{\substack{z \\ \mathbf{P}(Z=z)>0}} \mathbf{P}(Y=y, Z=z) H(X \mid Y=y, Z=z)$$

$$= \sum_{\substack{y,z \\ \mathbf{P}(Y=y,Z=z)>0}} \mathbf{P}(Y=y, Z=z) H(X \mid Y=y, Z=z)$$

$$= H(X \mid Y, Z).$$

This concludes the proof of the corollary. □

Remark 1.2.5. — A simpler proof is possible when the entropy $H(Y, Z)$ is finite. In this case, $H(Z)$ is finite too, and we have

$$\begin{aligned} H(X, Y \mid Z) &= H(X, Y, Z) - H(Z) \\ &= \big(H(X, Y, Z) - H(Y, Z)\big) + \big(H(Y, Z) - H(Z)\big) \\ &= H(X \mid Y, Z) + H(Y \mid Z). \end{aligned}$$

Corollary 1.2.6. — *Let $X_1, \ldots, X_n$ be discrete random variables; we have*

$$H(X_1, \ldots, X_n) = \sum_{k=1}^n H(X_k \mid X_1, \ldots, X_{k-1}).$$

Proof. — This follows from the previous corollary by induction on n:

$$\begin{aligned} H(X_1, \ldots, X_n) &= H(X_1) + H(X_2, \ldots, X_n \mid X_1) \\ &= H(X_1) + H(X_2 \mid X_1) + H(X_3, \ldots, X_n \mid X_1, X_2) \\ &= \cdots \\ &= H(X_1) + H(X_2 \mid X_1) + H(X_3 \mid X_1, X_2) + \cdots \\ &\quad + H(X_n \mid X_1, X_2, \ldots, X_{n-1}), \end{aligned}$$

as was to be shown. □

Remark 1.2.7. — The entropy of a random variable with values in a finite set A is a continuous function of its law. In the case of a uniform random variable, it is equal to $\log(\operatorname{Card}(A))$, hence is an increasing function of $\operatorname{Card}(A)$. Finally, we have the chain rule (proposition 1.2.3), as well as the relation $H(X \mid Y) = H(X)$ when the random values X and Y are independent. Conversely, let us show, after SHANNON (1948), that these properties characterize the entropy function, up to a normalizing factor.

Let $f(n)$ be the entropy of a uniform random variable on an n-element set. If $X_1, \ldots, X_k$ are independent uniform random variables on an n-element set, then $(X_1, \ldots, X_k)$ is a uniform random variable on a set of cardinality n^k. By the chain rule, we have $f(n^k) = k f(n)$. Let m and n be integers $\geqslant 2$ and let q be an integer; let $p = \lfloor q \log(n)/\log(m) \rfloor$ be the unique integer such that $m^p \leqslant n^q < m^{p+1}$. Since f is increasing, we have $f(m^p) \leqslant f(n^q) \leqslant f(m^{p+1})$, hence

$$pf(m) \leqslant qf(n) \leqslant (p+1)f(m).$$

When q tends to $+\infty$, p/q tends to $\log(n)/\log(m)$, as well as $(p+1)/q$, so that we get

$$f(m)\frac{\log(n)}{\log(m)} \leqslant f(n) \leqslant f(m)\frac{\log(n)}{\log(m)}.$$

We see that $f(n)/\log(n)$ is constant; in other words, there exists a real number c such that

$$f(n) = c\log(n)$$

for all n. Since f is increasing, we have $c > 0$.

Let now X be a discrete random variable on a finite set A; assume that the probabilities $p_a = \mathbf{P}(X = a)$ are rational numbers and write $p_a = n_a/N$, for some integers $(n_a)_{a \in A}$ and N. Applying the chain rule for each a, we may replace a by n_a possible values, each with probability $1/N$. The obtained random variable X^* is uniform on an N-element set, hence $H(X^*) = c\log(N)$. On the other hand, the chain rule implies that

$$H(X^*) = H(X) + \sum_{a \in A} p_a f(n_a) = H(X) + \sum_{a \in A} p_a c\log(n_a).$$

Therefore,

$$H(X) = c\log(N) - c\sum_{a \in A} p_a \log(n_a) = -c\sum_{a \in A} p_a \log(n_a/N)$$
$$= -c\sum_{a \in A} p_a \log(p_a).$$

In the general case, the continuity property of the entropy allows us to approximate the probabilities p_a by rational numbers, so that the preceding relation is still satisfied. This furnishes a justification for the a priori definition of entropy.

1.3. Mutual information

Definition 1.3.1. — *Let p, q be laws of discrete random variables on a set* A. *One calls the quantity*

$$D(p,q) = \sum_{\substack{a \in A \\ p(a) > 0}} p(a)\log\left(\frac{p(a)}{q(a)}\right)$$

the *divergence of q with respect to p.*

In fact, nothing assures us that this family is summable, and if there exists an element $a \in A$ such that $p(a) > 0$ and $q(a) = 0$, we get $D(p,q) = +\infty$. However, we shall check that the family

1.3. Mutual information.

$$\Big(p(a)\inf\big(\log(p(a)/q(a)),0\big)\Big)$$

is summable. This allows us to rewrite the divergence as the sum

$$D(p,q) = \sum_{\substack{a\in A \\ p(a)>0}} p(a)\inf\Big(\log\Big(\frac{p(a)}{q(a)}\Big),0\Big)$$

$$+ \sum_{\substack{a\in A \\ p(a)>0}} p(a)\Big(\log\Big(\frac{p(a)}{q(a)}\Big) - \inf\Big(\log\Big(\frac{p(a)}{q(a)}\Big),0\Big)\Big)$$

of two sums of families. The first one is finite, and the second has positive real terms hence is well defined as an element of $[0; +\infty]$, which shows that $D(p, q)$ is well defined as an element of $\mathbf{R} \cup \{+\infty\}$. Actually, the following theorem shows that $D(p, q)$ is a well-defined element of the interval $[0; +\infty]$.

Theorem 1.3.2. — *Let p, q be discrete laws on a set A.*

(i) The family $\Big(p(a)\inf\big(\log(p(a)/q(a)),0\big)\Big)$ is summable.
(ii) One has $D(p, q) \geqslant 0$, and equality holds if and only if $p = q$.

Proof. —
The logarithm function is strictly concave; its graph is below any of its tangent lines, and only meets this tangent line at that point. In particular, we have $\log(x) \geqslant x - 1$ for every $x \in \mathbf{R}_{>0}$ (an inequality that can also be proved by analysing the difference $\log(x) - x + 1$, or by the Taylor formula), and the inequality is strict for $x \neq 1$. Let us record the variant

$$\log \frac{1}{x} = -\log x \geqslant 1 - x,$$

with equality if and only if $x = 1$, and let us apply it to $x = q(a)/p(a)$, for any $a \in A$ such that $p(a) > 0$. We obtain

$$\log \frac{p(a)}{q(a)} \geqslant 1 - \frac{q(a)}{p(a)} = \frac{p(a) - q(a)}{p(a)},$$

hence

$$p(a)\log \frac{p(a)}{q(a)} \geqslant p(a) - q(a),$$

with equality if and only if $q(q) = p(a) > 0$. Summing over the set of all $a \in A$ such that $p(a) > 0$, we get

$$D(p,q) \geqslant 1 - \sum_{\substack{a\in A \\ p(a)>0}} q(a) \geqslant 0.$$

Let us assume that we have the equality $D(p, q) = 0$. Then $q(a) = p(a)$ for every a such that $p(a) > 0$, hence

$$1 = \sum_{a \in A} q(a) = \sum_{p(a)=0} q(a) + \sum_{p(a)>0} q(a)$$
$$= \sum_{p(a)=0} q(a) + \sum_{p(a)>0} p(a) = \sum_{p(a)=0} q(a) + 1,$$

so that $q(a) = 0$ as soon as $p(a) = 0$. This implies that $p = q$. Conversely, if $p = q$, the definition of the divergence immediately implies that $D(p, q) = 0$. □

Remark 1.3.3. — In the literature, the expression $D(p, q)$ is referred to as the *Kullback–Leibler divergence*, or the Kullback–Leibler distance: it measures the discrepancy between the two discrete laws p and q: it is positive, and only vanishes for $p = q$. However, it is not actually a distance, because it is not symmetric, and does not satisfy the triangle inequality. In fact, other expressions have analogous properties, and may serve similar needs; since we shall not use these more general functions in this introductory book, we prefer to simply refer to it as *divergence*. Let us also mention that its customary notation is $D(p \parallel q)$.

When X and Y are discrete random variables on a set A, with respective laws p and q, we shall also write $D(X, Y)$ for $D(p, q)$.

1.3.4. — Let X and Y be discrete random variables. On the set of all possible values of the pair (X, Y), we can consider the following two discrete laws.

1. The law of the pair (X, Y), given by $(x, y) \mapsto \mathbf{P}(X = x, Y = y)$;
2. The product of the two "marginal laws" of this pair, that is, $(x, y) \mapsto \mathbf{P}(X = x)\mathbf{P}(Y = y)$.

Definition 1.3.5. — Let X, Y be discrete random variables. We call the mutual information *of X and Y the divergence of the law* $(x, y) \mapsto \mathbf{P}(X = x)\mathbf{P}(Y = y)$, *the product of the two marginal laws of the pair* (X, Y), *with respect to the law of the pair* (X, Y). *We denote it by* $I(X, Y)$.

Explicitly, we thus have

$$I(X, Y) = \sum_{x,y} \mathbf{P}(X = x \text{ and } Y = y) \log \left(\frac{\mathbf{P}(X = x \text{ and } Y = y)}{\mathbf{P}(X = x)\mathbf{P}(Y = y)} \right).$$

Corollary 1.3.6. — *Let X and Y be discrete random variables. Their mutual information* $I(X, Y)$ *belongs to* $[0; +\infty]$; *it vanishes if and only if X and Y are independent.*

Proof. — The inequality $I(X, Y) \geq 0$ is a particular case of the theorem. By that theorem, we also know that $I(X, Y) = 0$ if and only if

$$\mathbf{P}(X = x, Y = y) = \mathbf{P}(X = x)\mathbf{P}(Y = y)$$

for every pair (x, y), which means precisely that X and Y are independent. □

1.3. Mutual information.

Corollary 1.3.7. — *Let X and Y be discrete random variables. The following equalities hold:*

$$H(X) = I(X, Y) + H(X \mid Y) \tag{1.3.7.1}$$
$$H(X, Y) + I(X, Y) = H(X) + H(Y). \tag{1.3.7.2}$$

In particular, we have the inequality

$$H(X \mid Y) \leqslant H(X). \tag{1.3.7.3}$$

In the case where the entropy of X is finite, equality $H(X) = H(X \mid Y)$ holds if and only if X and Y are independent.

If the entropy of a random variable measures its uncertainty, conditioning it to another random variable lowers this uncertainty.

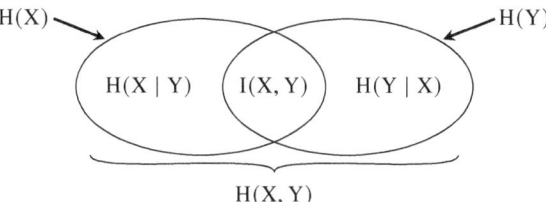

Fig. 1.5 Entropy of two random variables and of their pair, mutual information, conditional entropies

Proof. — When all these quantities are finite, we may use proposition 1.2.3 and write

$$H(X) - H(X \mid Y) = H(X) + H(Y) - H(X, Y)$$
$$= \sum_{x,y} \sum_x P(X = x, Y = y) \log \left(\frac{P(X = x)P(Y = y)}{P(X = x, Y = y)} \right)$$
$$= I(X, Y),$$

which gives the first relation.

In the general case, the fact that these terms might be infinite forces us to redo part of the computation that we did in the proof of proposition 1.2.3. Let us reuse the notation introduced there, setting $p_x = \mathbf{P}(X = x)$, $q_y = \mathbf{P}(Y = y)$ and $r_{xy} = \mathbf{P}(X = x \text{ and } Y = y)$. Going back to the definition of the mutual information $I(X, Y)$ and the conditional entropy $H(X \mid Y)$, we may write,

$$I(X, Y) + H(X \mid Y) = \sum_{x,y} r_{xy} \log \frac{r_{xy}}{p_x q_y} + \sum_y q_y H(X \mid Y = y)$$

$$= \sum_{x,y} r_{xy} \log \frac{r_{xy}}{p_x q_y} + \sum_y q_y \sum_x -\frac{r_{xy}}{q_y} \log \frac{r_{xy}}{q_y}$$

$$= \sum_{x,y} r_{xy} \log \frac{r_{xy}}{p_x q_y} + \sum_{x,y} q_y \sum_x -\frac{r_{xy}}{q_y} \log \frac{r_{xy}}{q_y}$$

$$= \sum_{x,y} r_{xy} \log \frac{r_{xy}}{p_x q_y} + \sum_{x,y} r_{xy} \log \frac{q_y}{r_{xy}}$$

$$= \sum_{x,y} r_{xy} \log \frac{1}{p_x} = \sum_x p_y \log \frac{1}{p_x} = H(X).$$

Let us add $H(Y)$ to both sides of this relation, to obtain a second one:

$$H(X, Y) + I(X, Y) = H(Y) + H(X \mid Y) + I(X, Y) = H(Y) + H(X).$$

Since $I(X, Y) \geqslant 0$, the last inequality also results from the first relation, as well as the equality case when the entropy of X is finite. Indeed, since $I(X, Y)$ and $H(X \mid Y)$ belong to $[0; +\infty]$, and since their sum is equal to $H(X)$, these quantities are all finite; equality $H(X) = H(X \mid Y)$ is then equivalent to $I(X, Y) = 0$, that is, to the independence of X and Y by corollary 1.3.6. □

Definition 1.3.8. — *Let* X, Y, Z *be discrete random variables. One says that* X *and* Z *are* conditionally independent *with respect to* Y, *and we write* $X \perp_Y Z$, *if we have*

$$\mathbf{P}(X = x, Z = z \mid Y = y) = \mathbf{P}(X = x \mid Y = y)\mathbf{P}(Z = z \mid Y = y),$$

for every x, y, z *such that* $\mathbf{P}(Y = y) > 0$.

Another notation, common in the theory of Markov chains, is $X \to Y \to Z$.

Example 1.3.9. — *Assume that there exists a function* f *such that* $Z = f(Y)$; *then* $X \perp_Y Z$. *In particular,* X *and* Y *are conditionally independent with respect to* Y.
Indeed, let x, y, z be such that $\mathbf{P}(Y = y) > 0$. If $z \neq f(y)$, then

$$\mathbf{P}(X = x, Z = z \mid Y = y) = 0 \quad \text{and} \quad \mathbf{P}(Z = z \mid Y = y) = 0.$$

On the other hand, if $z = f(y)$, then $\mathbf{P}(X = x, Z = z \mid Y = y) = \mathbf{P}(X = x \mid Y = y)$ and $\mathbf{P}(Z = z \mid Y = y) = 1$. In both cases, the desired inequality

$$\mathbf{P}(X = x, Z = z \mid Y = y) = \mathbf{P}(X = x \mid Y = y)\mathbf{P}(Z = z \mid Y = y)$$

holds.

1.3.10. — To study the notion of conditional independence of X, Z with respect to Y, it is useful to introduce the notion of *conditional mutual information of* X, Z *with respect to* Y, defined by

$$I(X, Z \mid Y) = \sum_y \mathbf{P}(Y = y) I(X \mid Y = y, Z \mid Y = y).$$

1.3. Mutual information.

It belongs to $[0;+\infty]$, and vanishes if and only if $I(X \mid Y = y, Z \mid Y = y) = 0$ for all y such that $\mathbf{P}(Y = y) > 0$, that is, if X and Z are conditionally independent with respect to Y. Moreover, to simplify our notation, let us write $p(x)$ for $\mathbf{P}(X = x)$, $p(x, y)$ for $\mathbf{P}(X = x, Y = y)$, $p(x, z \mid y)$ for $\mathbf{P}(X = x, Z = z \mid Y = y)$, etc. We then have

$$\begin{aligned} I\big(X, (Y, Z)\big) &= \sum_{x,y,z} p(x, y, z) \log \frac{p(x, y, z)}{p(x) p(y, z)} \\ &= \sum_{x,y,z} p(x, y, z) \log \frac{p(x, y, z) p(y)}{p(x, y) p(y, z)} + \sum_{x,y,z} p(x, y, z) \log \frac{p(x, y)}{p(x) p(y)} \\ &= \sum_y p(y) \Big(\sum_{x,z} p(x, z \mid y) \log \frac{p(x, z \mid y)}{p(x \mid y) p(z \mid y)} \Big) \\ &\quad + \sum_{x,y} p(x, y) \log \frac{p(x, y)}{p(x) p(y)} \\ &= I(X, Z \mid Y) + I(X, Y). \end{aligned}$$

By symmetry, we also have

$$I\big(X, (Y, Z)\big) = I(X, Y \mid Z) + I(X, Z),$$

so that $I(X, Y \mid Z) \geq 0$, and $I\big(X, (Y, Z)\big) \geq I(X, Z)$.

When X and Z are conditionally independent with respect to Y, one has $I(X, Z \mid Y) = 0$. These two expressions for $I\big(X, (Y, Z)\big)$ then imply the *data processing inequality* that appears in the following theorem.

Theorem 1.3.11. — *Let X, Y, Z be three discrete random variables. If* $X \perp_Y Z$, *then* $I(X, Y) \geq I(X, Z)$, *and equality holds if and only if* $X \perp_Z Y$.

Proof. — Since X and Z are assumed to be conditionally independent with respect to Y, we have $I(X, Z \mid Y) = 0$, hence

$$I\big(X, (Y, Z)\big) = I(X, Z \mid Y) + I(X, Y) = I(X, Y).$$

Therefore, we have

$$I(X, Y) - I(X, Z) = I\big(X, (Y, Z)\big) - I(X, Z) = I(X, Y \mid Z).$$

By definition of the conditional mutual information $I(X, Y \mid Z)$, this implies the data processing inequality $I(X, Y) \geq I(X, Z)$, as well as the characterization of the equality case. □

Corollary 1.3.12. — *Let X, Y be discrete random variables and let f be a function. One has* $I\big(X, f(Y)\big) \leq I(X, Y)$, *and equality holds if and only if* $X \perp_{f(Y)} Y$.

Proof. — Set $Z = f(Y)$. We saw in example 1.3.9 that X and Z are conditionally independent with respect to Y. By theorem 1.3.11, we thus have

$$I(X, f(Y)) = I(X, Z) \geq I(X, Y),$$

and the equality case is proved similarly. □

1.4. Entropy rate

In information theory, random variables do not represent messages (a text, a photograph, a sound) but rather the individual elements (successive letters, pixels and their color, etc.) that constitute this message. In the previous section, we have learnt to treat several random variables as a single one, grouping them as a vector, but this method makes us lose the idea that it is the succession of a large number of elementary symbols.

Therefore, we will now consider infinite sequences (X_n), indexed by the natural numbers, of random variables—something we call a *stochastic process*.

Definition 1.4.1. — *The* entropy rate *of a stochastic process* $X = (X_n)$ *is the expression*

$$H(X) = \lim_{n \to \infty} \frac{1}{n+1} H(X_0, \ldots, X_n),$$

provided the limit exists.

By replacing the limit by upper and lower limits, we define analogously the upper entropy rate $\overline{H}(X)$, and the lower entropy rate $\underline{H}(X)$, which have the advantage of always being defined. By definition, they satisfy the inequality $\underline{H}(X) \leq \overline{H}(X)$, and the entropy rate exists if and only if the lower and upper entropy rates coincide, to which it is then equal.

Example 1.4.2. — Let $X = (X_n)$ be a stochastic process. Assume that the random variables X_n are independent. By the chain rule, one then has

$$\frac{1}{n+1} H(X_0, \ldots, X_n) = \frac{1}{n+1} \sum_{k=0}^{n} H(X_k),$$

hence the entropy rate of this stochastic process is the Cesàro limit of the sequence $(H(X_n))$. Let us moreover assume that the random variables X_n have the same law. Then $H(X_k) = H(X_0)$ for all k and one has $H(X) = H(X_0)$.

Lemma 1.4.3 (Cesàro [4]). — *Let (a_n) be a sequence of real numbers; for every integer $n \geq 0$, set $A_n = (a_0 + \cdots + a_n)/(n+1)$. We have the inequalities*

$$\underline{\lim} \, a_n \leq \underline{\lim} \, A_n \leq \overline{\lim} \, A_n \leq \overline{\lim} \, a_n.$$

---

[4] Ernesto Cesàro (1859–1906) was an Italian mathematician, known for his work in differential geometry. In particular, he defined and studied self-similar curves ("fractals"), such as von Koch's snowflake curve.

1.4. Entropy rate.

In particular, if the sequence (a_n) *admits a limit* $\ell \in [-\infty; +\infty]$, *then the sequence* (A_n) *converges to* ℓ *too.*

Proof. — Let us prove the inequality $\overline{\lim} A_n \leq \overline{\lim} a_n$.

There is nothing to prove when $\overline{\lim} a_n = +\infty$; let us therefore assume that $\overline{\lim} a_n < \infty$, and let λ be a real number such that $\overline{\lim} a_n < \lambda$. By definition of the upper limit, there exists an integer N such that $a_n \leq \lambda$ for every integer $n \geq N$. For $n \geq N$, we then have

$$A_n = \frac{1}{n+1} \sum_{k=0}^{n} a_k$$
$$= \frac{1}{n+1} \sum_{k=0}^{N-1} a_k + \frac{1}{n+1} \sum_{k=N}^{n} a_k$$
$$\leq \frac{1}{n+1} \sum_{k=0}^{N-1} a_k + \frac{n+1-N}{n+1} \lambda.$$

When n goes to infinity, the right-hand side of this inequality converges to λ. This implies in particular that $\overline{\lim} A_n \leq \lambda$. Since λ is arbitrary, this shows the inequality $\overline{\lim} A_n \leq \overline{\lim} a_n$.

By replacing the sequence (a_n) by the sequence (b_n) defined by $b_n = -a_n$, the sequence (A_n) is replaced by the sequence (B_n) defined by $B_n = -A_n$. We then have $\underline{\lim} a_n = -\overline{\lim} b_n$ and $\underline{\lim} A_n = -\overline{\lim} B_n$. Applied to the sequence (b_n), the inequality between upper limits implies that $\underline{\lim} a_n \leq \underline{\lim} A_n$.

When the sequence (a_n) converges to an element ℓ in $[-\infty; +\infty]$, one has $\underline{\lim} a_n = \ell = \overline{\lim} a_n$, and the above inequalities imply that $\underline{\lim} A_n = \overline{\lim} A_n = \ell$, so that the sequence (A_n) converges to ℓ. □

Definition 1.4.4. — *One says that a stochastic process* (X_n) *is stationary if for every sequence* $(x_0, \ldots, x_m)$, *we have*

$$\mathbf{P}(X_n = x_0, X_{n+1} = x_1, \ldots, X_{n+m} = x_m) = \mathbf{P}(X_0 = x_0, X_1 = x_1, \ldots, X_m = x_m)$$

for all $n \in \mathbf{N}$.

Taking $m = 0$, we see that the terms of a stationary stochastic process have the same law.

Proposition 1.4.5. — *Let* $X = (X_n)$ *be a stationary stochastic process. Then the entropy rate* $H(X)$ *exists, and is given by the formula*

$$H(X) = \lim_{n \to \infty} H(X_n \mid X_{n-1}, \ldots, X_0).$$

Proof. — For every integer n, let us put

$$H'(X)_n = H(X_n \mid X_{n-1}, \ldots, X_0).$$

Since the entropy decreases by conditioning, we have

$$H'(X)_{n+1} = H(X_{n+1} \mid X_n, \ldots, X_0) \leqslant H(X_{n+1} \mid X_n, \ldots, X_1)$$

for every n. Since the stochastic process X is stationary, we have

$$H(X_{n+1} \mid X_n, \ldots, X_1) = H(X_n \mid X_{n-1}, \ldots, X_0) = H'(X)_n.$$

Consequently, the sequence $\bigl(H'(X)\bigr)_n$ is decreasing. Since it is positive, it converges to an element of $[0; +\infty]$; let us denote it by $H'(X)$.

Then, for every integer n, we have

$$H(X_0, \ldots, X_n) = H(X_0) + H(X_1 \mid X_0) + \cdots + H(X_n \mid X_{n-1}, \ldots, X_0)$$
$$= \sum_{k=0}^{n} H'(X)_k,$$

which proves that the sequence $\left(\frac{1}{n+1} H(X_0, \ldots, X_n)\right)$ is the sequence of Cesàro means of $(H'(X)_n)$. It thus converges to its limit, as was to be shown. □

1.5. Entropy rate of Markov processes

1.5.1. — Two basic assumptions regarding the random variables of a stochastic process allowed us to study its entropy rate: independence and stationarity. However, these two assumptions are not very well adapted to applications. Independence contradicts the very idea that these variables constitute a message that makes sense; similarly, if a symbol is possible, then any doubling, tripling, etc. of that symbol will be possible, which contradicts the absence of words in which the letter "t" is used ten times, for example, or the observation that in English the letter "q" is almost always followed by a "u". As for the stationarity hypothesis, it neglects the observation that the beginning or the end of a message, or the edge of an image are of a different nature from the core of the message.

The markovian hypothesis that we now introduce is already closer to reality. In an intuitive way, it consists in saying that a random variable X_{n+1} can depend on the previous ones, but not more than through the variable X_n. This is the model for a random walk in a city. Assuming that the X_n represent letters, a simple example already shows its limits: by allowing the succession *tt*, we are forced to allow the tripling *ttt*, etc.

1.5. Entropy rate of Markov processes.

Definition 1.5.2. — *We say that a stochastic process* (X_n) *is* markovian *(or is a* Markov[5] *process, or is a* Markov chain*) if for any integer* $n \geq 1$, $(X_0, \ldots, X_{n-1})$ *and* X_{n+1} *are conditionally independent with respect to* X_n.

This means that for any integer n and any sequence $(x_0, \ldots, x_{n+1})$, we have

$$\mathbf{P}(X_{n+1} = x_{n+1} \mid X_n = x_n, \ldots, X_0 = x_0) = \mathbf{P}(X_{n+1} = x_{n+1} \mid X_n = x_n).$$

1.5.3. — Let $X = (X_n)$ be a Markov process. We make the additional hypothesis that it is *homogeneous*, that is to say that for any pair (a, b), we have

$$\mathbf{P}(X_{n+1} = b \mid X_n = a) = \mathbf{P}(X_1 = b \mid X_0 = a).$$

Let us suppose that the set A of possible values of (X_n) is finite; for any pair $(a, b) \in A^2$, let $p_{a,b} = \mathbf{P}(X_1 = b \mid X_0 = a)$ and let us introduce the matrix $P = (p_{a,b})$. It is a square matrix with indices in the set A. Most often in linear algebra, rows and columns of matrices are indexed by the successive integers $1, 2, \ldots$, which corresponds to the way we index bases of vector spaces, but the theory is identical (possibly clearer, in fact) if we index them by arbitrary finite sets. The coefficients of the matrix P are conditional probabilities; they are therefore positive. For all a, we have

$$\sum_{b \in A} p_{a,b} = \sum_{b \in A} \mathbf{P}(X_1 = b \mid X_0 = a) = 1.$$

In other words, the sum of the coefficients of each row of P is equal to 1. We say that P is a *stochastic matrix*.

The matrix P is called the transition matrix of the Markov process X

In the vocabulary of Markov chains, the elements of A are called *states*, and $p_{a,b}$ is the probability of passing from state a to state b. We often represent such a chain by a *quiver* (or *directed graph*) whose *vertices* are the states of the chain, endowed, for each pair of states (a, b) such that $p_{a,b} > 0$, with an arrow from state a to state b labeled with probability $p_{a,b}$.

Thus, the quiver of figure 1.6 represents a Markov chain with two states $\{a, b\}$ for which the probability of going from a to b is equal to p, and the probability of going from b to a is equal to q. Its transition matrix is thus given by

$$P = \begin{pmatrix} 1-p & p \\ q & 1-q \end{pmatrix}.$$

If $M = (\mu_a)$ is the law of X_n, considered as a row-vector, then the law $M' = (\mu'_a)$ of X_{n+1} is given by

$$\mu'_a = \mathbf{P}(X_{n+1} = a) = \sum_{b \in A} \mathbf{P}(X_n = b) \times \mathbf{P}(X_{n+1} = a \mid X_n = b) = \sum_{b \in A} \mu_b p_{b,a}.$$

---

[5] Andrey MARKOV (1856–1922) was a Russian mathematician whose name has remained attached to several notions of probability (Markov chain here, but also the Markov inequality), as well as to the study of the diophantine equation $x^2 + y^2 + z^2 = 3xyz$.

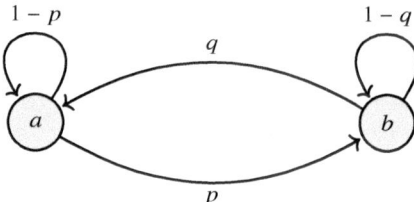

Fig. 1.6 A two-state Markov chain

In other words, we have the matrix relation

$$M' = M \cdot P.$$

By induction, if the vector M represents the law of X_0, the law of X_n is represented by the vector MP^n.

The following proposition characterizes the stationary Markov processes and computes their entropy rate as a function of their transition matrix.

Proposition 1.5.4. — *Let $X = (X_n)$ be a homogeneous Markov process. Let A be the set of values of X_0, let $P = (p_{a,b})$ be the transition matrix of X and let $M = (\mu_a)$ be the law of X_0. For the process X to be stationary, it is necessary and sufficient that we have $M = MP$. In this case, we have*

$$H(X) = -\sum_{a,b} \mu_a p_{a,b} \log(p_{a,b}).$$

We say that the law of X_0 (or that the vector M) is a *stationary law* for this Markov process.

Proof. — If X is stationary, then X_0 and X_1 have the same law, hence $M = MP$. Let us suppose, conversely, that $M = MP$ and let us prove that X is a stationary process. By induction, we have $MP^n = M$ for any integer n, so that the law of X_n is given by $MP^n = M$.

Let us prove by induction on m that:

$$P(X_n = x_0, \ldots, X_{n+m} = x_m) = P(X_0 = x_0, \ldots, X_m = x_m),$$

for any integers m, n and any states $x_0, \ldots, x_m$. By definition of a homogeneous Markov process, we have

$$\begin{aligned}
P(X_n &= x_0, \ldots, X_{n+m} = x_m) \\
&= P(X_{n+m} = x_m \mid X_n = x_0, \ldots, X_{n+m-1} = x_{m-1}) \\
&\quad \times P(X_n = x_0, \ldots, X_{n+m-1} = x_{m-1}) \\
&= P(X_{n+m} = x_m \mid X_{n+m-1} = x_{m-1}) \times P(X_n = x_0, \ldots, X_{n+m-1} = x_{m-1}) \\
&= P(X_1 = x_m \mid X_0 = x_{m-1}) \times P(X_n = x_0, \ldots, X_{n+m-1} = x_{m-1}).
\end{aligned}$$

1.5. Entropy rate of Markov processes.

By the induction hypothesis, we have

$$P(X_n = x_0, \ldots, X_{n+m-1} = x_{m-1}) = P(X_0 = x_0, \ldots, X_{m-1} = x_{m-1}).$$

Therefore,

$$\begin{aligned}
P(X_n &= x_0, \ldots, X_{n+m} = x_m) \\
&= P(X_1 = x_m \mid X_0 = x_{m-1}) \times P(X_0 = x_0, \ldots, X_{m-1} = x_{m-1}) \\
&= P(X_0 = x_0, \ldots, X_{m-1} = x_{m-1}, X_m = x_m),
\end{aligned}$$

using once again the hypothesis that the process X is markovian. This proves that X is a stationary process.

Let us now assume that the stochastic process X is stationary and prove the given formula for its entropy rate. For any integer n, we have $H(X_n \mid X_{n-1}, \ldots, X_0) = H(X_n \mid X_{n-1})$, by the Markov property, then $H(X_n \mid X_{n-1}) = H(X_1 \mid X_0)$ by stationarity. We thus have $H(X_n \mid X_{n-1}, \ldots, X_0) = H(X_1 \mid X_0)$, hence the formula $H(X) = H(X_1 \mid X_0)$ by proposition 1.4.5. Moreover, the definition of conditional entropy leads to

$$H(X_1 \mid X_0) = \sum_a P(X_0 = a) H(X_1 \mid X_0 = a) = -\sum_a \mu_a \sum_b p_{a,b} \log(p_{a,b}),$$

as was to be shown. □

Example 1.5.5. — Let us go back to the example of the two-state Markov chain X represented by the graph of figure 1.6. The laws of X_0 for which this chain is stationary are given by a vector (μ, ν) such that

$$\begin{cases} \mu(1-p) + \nu q = \mu \\ \mu p + nu(1-q) = \nu. \end{cases}$$

We obtain the equality $p\mu = q\nu$. Joined to the conditions $\mu + \nu = 1$ and $\mu, \nu \geq 0$, we see that there exists a unique such law, given by

$$(\mu, \nu) = \left(\frac{q}{p+q}, \frac{p}{p+q} \right).$$

Then, the entropy rate of X is equal to

$$\begin{aligned}
H(X) &= \frac{q}{p+q} \left(-(1-p)\log(1-p) - p\log(p) \right) \\
&\quad + \frac{p}{p+q} \left(-q\log(q) - (1-q)\log(1-q) \right) \\
&= \frac{q}{p+q} h(p) + \frac{p}{p+q} h(q),
\end{aligned}$$

where h is the entropy function represented in figure 1.4, page 27.

Moreover, for any integer n, the law of X_n is again given by (μ, ν), so that the entropy of X_n is equal to

$$H(X_n) = -\frac{q}{p+q}\log\left(\frac{q}{p+q}\right) - \frac{p}{p+q}\log\left(\frac{p}{p+q}\right) = h\left(\frac{p}{p+q}\right).$$

One can ask what values of p and q make these expressions extremal. Recall that the function h is zero at 0 and 1, strictly increasing on $[0; 1/2]$ and strictly decreasing on $[1/2; 1]$. Thus, $H(X_n) = 0$ if, and only if, $p/(p+q) = 0$ or 1, that is, $p = 0$ or $q = 0$; and, $H(X_n)$ is maximal when $p/(p+q) = 1/2$, i.e. $p = q$.

The function h is concave. We therefore have the following inequality:

$$H(X) = \frac{q}{q+p}h(p) + \frac{p}{p+q}h(q) \geq h\left(\frac{q}{q+p}p + \frac{p}{p+q}q\right) = h\left(\frac{2pq}{p+q}\right),$$

with equality if and only if $p = q$. On the other hand, $h(2pq/(p+q))$ is maximal when $2pq/(p+q) = 1/2$, that is $4pq = p+q$. Finally, we find that $H(X)$ is maximal if and only if $p = q = 1/2$.

Definition 1.5.6. — *Let $X = (X_n)$ be a homogeneous Markov chain with finite set of states A, and let $P = (p_{a,b})$ be its transition matrix. We say that the chain X, or that the stochastic matrix P, is* primitive *if there exists an integer $m \geq 1$ such that all the coefficients of the matrix P^m are strictly positive.*

Theorem 1.5.7 (O. Perron [6], 1907). — *Let P be a primitive stochastic matrix. The sequence of matrices (P^n) converges, and its limit Q is a stochastic matrix of rank 1.*

Proof. — We provide the space $\mathbf{R}^A$ with the norm defined by $\|X\| = \sum_{a \in A}|x_a|$, for $X = (x_a)$; let $f(X) = \sum_{a \in A} x_a$. We also denote by Σ the set of $X \in \mathbf{R}^A_+$ such that $f(X) = 1$; it is a simplex in $\mathbf{R}^A$ (a segment, a triangle, a tetrahedron... when the number of states is equal to $2, 3, 4\ldots$) in particular, it is convex, compact and nonempty. Let m be an integer ≥ 1 such that all coefficients $(p'_{a,b})$ of the matrix $P' = P^m$ are strictly positive; let us denote by c their lower bound.

The proof of the theorem requires several steps.

a) For all $X \in \mathbf{R}^A$, we have $\|XP\| \leq \|X\|$ and $f(XP) = f(X)$.

Set $Y = XP$; then we have $Y = (y_b)$, with $y_b = \sum_a x_a p_{a,b}$. Thanks to the triangle inequality, we get

$$\|Y\| \sum_{b \in A}|y_b| \leq \sum_{a,b}|x_a|p_{a,b} = \sum_a |x_a|\sum_b p_{a,b} = \sum_a |x_a| = \|X\|.$$

In the same way,

---

[6] Oskar PERRON (1880–1975) was a German mathematician. His work concerned complex analysis, number theory and linear algebra. In 1912, this theorem by Perron was extended by G. Frobenius to the case of matrices which are not necessarily primitive; it had a great media success at the end of the 1990s because it was underlying the *PageRank* algorithm used by Google to evaluate the relevance of web pages.

1.5. Entropy rate of Markov processes.

$$f(Y) = \sum_b y_b = \sum_{a,b} x_a p_{a,b} = \sum_a x_a = f(X).$$

b) *For all $X \in \mathbf{R}_+^A$, the coefficients of XP are positive, and those of XP' are greater than $c\|X\|$.*

Let $X \in \mathbf{R}_+^A$, let $Y = XP$, and let $Y = (y_b)$. Then, we have $y_b = \sum_a x_a p_{a,b}$, so that $y_b \geq 0$ for all b.

Similarly, let $Z = XP' = (z_b)$. Since $p'_{a,b} \geq c$ and $x_a \geq 0$ for all a, b, we obtain

$$z_b = \sum_a x_a p'_{a,b} \geq \sum_a x_a c = c.$$

c) *Let $X \in \mathbf{R}^A$ be such that $f(X) = 0$. Then,*

$$\|XP'\| \leq \big(1 - c \operatorname{Card}(A)\big)\|X\|.$$

Let $Z = XP'$; write $Z = (z_b)$ so that $z_b = \sum_a x_a p'_{a,b}$. Since $\sum x_a = 0$ and $p'_{a,b} \geq c$ for all a, b, we also have $z_b = \sum_a x_a (p'_{a,b} - c)$, so that

$$|z_b| \leq \sum_a |x_a|(p'_{a,b} - c) \leq \sum_a |x_a| p'_{a,b} - c\|X\|.$$

Summing over b, we obtain

$$\|Z\| \leq \|X\| - c\operatorname{Card}(A)\|X\| = \big(1 - c\operatorname{Card}(A)\big)\|X\|.$$

d) *The set Σ is stable under the map $X \mapsto XP'$, and this map is contracting for the norm $\|\cdot\|$.*

The stability of Σ follows from what has been said above. Moreover, let X, X' be elements of Σ; let $Y = XP'$ and $Y' = X'P'$. We have $f(X' - X) = 0$ and $Y' - Y = (X' - X)P'$; then,

$$\|Y' - Y\| \leq \big(1 - c\operatorname{Card}(A)\big)\|X' - X\|,$$

hence the assertion since $1 - c\operatorname{Card}(A) < 1$.

e) *The sequence of matrices $((P')^n)$ converges; its limit is a stochastic matrix of rank 1.*

According to the Banach fixed point theorem (see lemma 1.5.9), the map $X \mapsto XP'$ has a single fixed point M in Σ and, for any vector $X \in \Sigma$, the sequence $(X(P')^n)$ converges to M.

The space Σ contains the vectors X_a of the canonical basis. For each of them, we have $X_a(P')^n \to M$. This proves that the a-th row of the sequence of matrices $((P')^n)$ converges to M. The sequence $((P')^n)$ therefore converges to the matrix Q, all of whose rows are equal to M; it is a stochastic matrix of rank 1.

f) *The sequence of matrices (P^n) converges to Q.*

Writing the Euclidean division of n by m, $n = mk + d$, where $0 \leq d \leq m - 1$, we have $P^n = P^d (P')^k$. Suppose that n tends to infinity while remaining in the congruence class modulo m of d; then we have $P^n \to P^d Q$. Now, since all rows of P^d belong to Σ, we have $P^d Q = Q$. Therefore, all of these subsequences converge to Q, which implies that the sequence (P^n) converges to Q. □

Remark 1.5.8. — All the rows of the matrix Q are equal to the same vector M with positive coefficients, of sum equal to 1. We have $MP = M$, which proves that M is a "left-eigenvector" of P, for the eigenvalue 1.

Let N be a left-eigenvector of P for an eigenvalue λ. Then $NP = \lambda N$ and, by induction, $NP^n = \lambda^n N$ for all n. When n tends to infinity, the left-hand side of this equality tends to NQ. Since $N \neq 0$, this implies that the sequence (λ^n) converges: we have therefore $\lambda = 1$ or $|\lambda| < 1$.

Let us assume that $\lambda = 1$. Let $X = N - f(N)M$, so that $f(X) = 0$. According to the point c) of the proof, we thus have

$$\|XQ\| \leq \big(1 - c\,\mathrm{Card}(A)\big)\|X\|.$$

Then $X = XP = XQ$, which implies $X = 0$ and $N = f(N)M$.

We have thus shown that under the conditions of the theorem, the matrix P has a single eigenvalue of modulus ≥ 1, this eigenvalue is equal to 1 and the corresponding "left-eigenspace" has dimension 1, generated by M.

For the sake of completeness, let us state and prove the fixed point theorem of Banach[7].

Lemma 1.5.9 (Banach fixed point theorem). — *Let* E *be a complete, non-empty metric space, and let* $f\colon \mathrm{E} \to \mathrm{E}$ *be a contracting map, in the sense that there exists a* $c < 1$ *such that* $d(f(x), f(y)) \leq c d(x, y)$ *for all* $x, y \in \mathrm{E}$, *where* d *is the distance of* E. *Then the map* f *has one, and only one, fixed point in* E.

Proof. — The assumption on f implies that it is continuous. It also implies that f cannot have more than one fixed point: if a, b are fixed points of f, we have $f(a) = a$ and $f(b) = b$ by definition, so $d(a, b) = d(f(a), f(b)) \leq c d(a, b)$, and the hypothesis $c < 1$ implies that $d(a, b) = 0$; in other words, $a = b$. To prove the existence of a fixed point, we use an iterative method. Let a_0 be any point of E. (this is where we use the hypothesis that E is not empty), and let (a_n) be the sequence in E defined by the recurrence relation $a_{n+1} = f(a_n)$. For $n \geq 1$, we have

$$d(a_n, a_{n+1}) = d(f(a_{n-1}), f(a_n)) \leq c d(a_{n-1}, a_n).$$

By induction, we deduce that $d(a_n, a_{n+1}) \leq c^n d(a_0, a_1)$ for all $n \geq 0$.

---

[7] Stefan BANACH (1892–1945) was a Polish mathematician. His work in topology and functional analysis made him one of the the greatest analysts of the 20th century. During the Second World War, he escaped the Nazi massacre of the professors of the University of Lwów (1941), but died of lung cancer at the end of the war.

1.5. Entropy rate of Markov processes.

Then, for $n, p \geq 0$, we have

$$d(a_n, a_{n+p}) \leq d(a_n, a_{n+1}) + dots + d(a_{n+p-1}, a_{n+p})$$
$$\leq c^n d(a_0, a_1) + dots + c^{n+p-1} d(a_0, a_1)$$
$$\leq \frac{c^n}{1-c} d(a_0, a_1).$$

The sequence (a_n) is thus a Cauchy sequence. Since the space E is complete, this sequence (a_n) has a limit a in E. We have $a_n \to a$, so $a_{n+1} = f(a_n) \to f(a)$, because f is continuous; therefore $a = f(a)$, which shows that a is a fixed point of f. □

Theorem 1.5.10. — *Let X be a homogeneous Markov chain, with finite set of states A, whose transition matrix $P = (p_{a,b})$ is primitive; let $M = (\mu_a)$ be its unique stationary law. Then, X has a rate of entropy, given by*

$$H(X) = \lim_{n \to \infty} H(X_n \mid X_{n-1}) = -\sum_{a,b} \mu_a p_{a,b} \log(p_{a,b}).$$

Proof. — Let $M_0 = (m_a^{(0)})$ be the vector of $\mathbf{R}^A$ describing the law of X_0. For any integer n, the law of X_n is represented by the vector $M_n = M_0 P^n = (m_a^{(n)})$. As a consequence,

$$H(X_n \mid X_{n-1}) = \sum_a m_a^{(n-1)} H(X_n \mid X_{n-1} = a) = -\sum_{a,b} m_a^{(n-1)} p_{a,b} \log(p_{a,b}).$$

Since $m_a^{(n)} \to \mu_a$ for all a, we have

$$H(X_n \mid X_{n-1}) \to -\sum_{a,b} \mu_a p_{a,b} \log(p_{a,b}).$$

Let us denote this expression by $H'(X)$.

Moreover, we have

$$H(X_1, \ldots, X_n) = H(X_1) + H(X_2 \mid X_1) + \cdots + H(X_n \mid X_{n-1}, \ldots, X_1).$$

Since X is a Markov process, we have the equality

$$H(X_n \mid X_{n-1}, \ldots, X_1) = H(X_n \mid X_{n-1}),$$

so that

$$H(X_1, \ldots, X_n) = H(X_1) + \sum_{k=2}^n H(X_k \mid X_{k-1}).$$

By Cesàro's lemma (lemma 1.4.3), we have

$$\frac{1}{n}H(X_1,\ldots,X_n) \to H'(X),$$

which proves that the entropy rate of X exists and is equal to $H'(X)$. □

Remark 1.5.11. — Let X be a homogeneous Markov chain with finite set of states A, and let $P = (p_{a,b})$ be its transition matrix. There is a convenient geometric way, using the quiver that represents this chain, to determine whether it is primitive or not. The coefficient $p_{a,b}^{(m)}$ of the matrix P^m is the probability of going from state a to state b in exactly m steps and we have

$$p_{a,b}^{(m)} = \sum_{a_1,\ldots,a_{m-1}\in A} p_{a,a_1} p_{a_1,a_2} \cdots p_{a_{m-1},b}.$$

Consequently, the coefficient $p_{a,b}^{(m)}$ is strictly positive if and only if there exists a sequence $(a_0, a_1, \ldots, a_m)$ of states such that $a_0 = a$, $a_m = b$, and such that $p_{a_i,a_j} > 0$ for all $i, j \in \{0,\ldots,m\}$, in other words, if and only if there exists a "path" of length m in the quiver associated to X which follows the arrows in their prescribed direction.

A necessary condition for the Markov chain X to be primitive is that, for any pair of states (a, b), there exists a path connecting vertex a to vertex b. If this holds, we say that this quiver is *connected* (or, in the terminology of directed graphs, that this directed graph is strongly connected), or that the chain is *irreducible*.

This condition is not sufficient: we need to forbid, for example, that the states can be colored with two colors so that no arrow connects states of the same color. In this case, any path between two states has even length if these states are of the same color, and odd length otherwise. Let us define the *period* of the Markov chain as the gcd of the lengths of the paths connecting a state to itself; we say that the Markov chain is *aperiodic* if this period is equal to 1.

We can then see that the Markov chain X is primitive if and only if it is irreducible and aperiodic.

Although we will not consider examples of Markov chains for which these hypotheses are not satisfied, it might be worth indicating in a few words how it is possible to reduce the general case to the situation where these conditions hold.

Say that two states $a, b \in A$ "communicate" if there exists a path from a to b as well as a path from b to a; this is an equivalence relation on A. Fix a "communication class" $C \subseteq A$ and consider only the edges of the initial quiver which link two states from that class; by construction, one obtains a quiver which is strongly connected. The Markov chain with set of states C that we get has for transition matrix a submatrix P_C, and it is irreducible.

All states of C have the same period, say d, and the class C naturally splits into d classes $C_1, \ldots, C_d$ such that the edges go from C_1 to C_2, C_2 to C_3, etc., and C_d to C_1. The stochastic matrix P_C^d is the transition matrix of a Markov chain with set of states C, and whose communication classes are $C_1, \ldots, C_d$. Each of them gives rise to a Markov chain which is irreducible and aperiodic, thanks to which the initial Markov chain can be studied.

Exercises

Exercise 1.1 (Function of a random variable, I). [p. 148] Amy likes to bet on horse races. Eight horses are competing. At the end of the race, her gain depends on the position reached by the horse she bet on: if it reaches the goal at position $1, 2, 3, \ldots, 8$, her gain is respectively

$$100, \ 50, \ 50, \ 0, \ -10, \ -50, \ -50, \ -100 \ \text{€}.$$

We will assume that all horses are of a comparable level.

a) Let X be the random variable describing Amy's gain. Compute its expectation $\mathbf{E}(X)$ and its base 2 entropy $H_2(X)$.

Since she is tired of losing money in the long term, Amy decides to monetize her horse-race playing by streaming her reactions on social networks. She makes a deal with some network so that her gains are now the *absolute value* of her original gains, since the public likes it as much when she wins as when she throws a tantrum because she loses.

b) Let $Y = |X|$ be the random variable corresponding to her gain. Compute its law. (*We recall that for any function f, the law of the random variable $f(X)$ is given by* $\mathbf{P}(f(X) = y) = \sum_{\{x \in \mathcal{X} | f(x) = y\}} \mathbf{P}(X = x)$.)

c) Compute $\mathbf{E}(Y)$ and $H_2(Y)$. What do you observe?

Exercise 1.2. [p. 149] An urn contains w white balls and r red balls. We denote by $X = (X_1, \ldots, X_5)$, resp. $Y = (Y_1, \ldots, Y_5)$, the random variable that describes the colors of the five balls picked up randomly from the urn, putting them back in the urn in the first case, and keeping them out in the second. Which of these two random variables has the greatest entropy? (*Prove that the variables $X_1, \ldots, X_5$ are independent and follow the same law; prove that the variables $Y_1, \ldots, Y_5$ also follow this law, but that they are not independent.*)

Exercise 1.3 (Uniform law). [p. 150] Let X be a uniform random variable on a finite set $\mathcal{X} = \{x_1, \ldots, x_n\}$. Let Y be another discrete random variable on $\mathcal{X}$.

a) Recall, for every $x \in \mathcal{X}$, the probability $p(x) = \mathbf{P}(X = x)$. What is the entropy $H(X)$?

b) Prove that the function $x \mapsto x \log(x)$ on $[0; 1]$ is strictly convex, and deduce that if $q_1, \ldots, q_n$ are real numbers in $[0; 1]$, with mean q, one has $q \log(q) \leq \frac{1}{n} \sum_{i=1}^{n} q_i \log(q_i)$, and that equality holds if and only if $q_i = q$.

c) Prove that the entropy of a random variable on $\mathcal{X}$ is always smaller than or equal to $\log(n)$, and that the uniform law is the unique one that achieves this maximum.

d) Let $q\colon \mathcal{X} \to [0;1]$ be the law of Y. Prove that $D(q, p) = H(X) - H(Y)$ and recover the result of the previous question.

Exercise 1.4 (Geometric law on N). [p. 150] We toss a coin many times until we obtain heads. Let X be the random variable with values in **N** that indicates the number of coin tosses we needed to make until heads appeared: 0 if we got heads on the first throw, 1 if it appeared on the second throw, etc. The coin is not assumed to be fair, and we denote by x the probability of getting heads at a given coin toss; we assume $0 < x < 1$.

a) Compute $\mathbf{P}(X = 0)$, $\mathbf{P}(X = 1)$, $\mathbf{P}(X = 2)$. Compute $\mathbf{P}(X = k)$ for all $k \in \mathbf{N}$.
b) Compute the expectation $\mathbf{E}(X)$.
You may use the relation $\sum_{k \in \mathbf{N}} kt^k = t/(1-t)^2$, for $|t| < 1$.
c) Show that $H(X) = -\log(x) - \frac{1-x}{x}\log(1-x)$.
d) Let $h(x)$ be the entropy of X when the probability of getting heads is x. Show that h is a decreasing function.
e) Compute the limit of $h(x)$ when $x \to 0$ and $x \to 1$. Does there exist a law of maximal entropy on **N**? (*Use the fact that* $\lim_{t \to 0} t \log t = 0$.)
f) Let Y be a discrete random variable with values in **N** such that $\mathbf{E}(Y) = \mathbf{E}(X)$. Let p and q be the laws of X and Y, respectively. Prove that $H(X) - H(Y) = D(q, p)$. (*Compare with exercise 1.3.*)
g) Prove that among all laws on **N** which have a given expectation $m > 0$, the geometric law with parameter $x = 1/(m+1)$ is the only one that maximizes entropy. Compute this maximal value of the entropy, as a function of m.

Exercise 1.5 (Function of a random variable, II). [p. 153] Let X be a discrete random variable with values in a set $\mathcal{X}$ and let p be its law. Let $\mathcal{Y}$ be a second set and let $f\colon \mathcal{X} \to \mathcal{Y}$ be any map. We denote by $Y = f(X)$ the discrete random variable, the image of X under f.

a) Using the definition of conditional entropy, prove that $H(f(X) \mid X) = 0$. Deduce that $H(X, f(X)) = H(X)$.
b) Prove that $H(X) \geqslant H(f(X))$.
c) Assuming that f is injective, prove that $H(X \mid f(X)) = 0$ and that $H(X) = H(f(X))$.
d) More generally, prove that $H(X) = H(f(X))$ if and only if the restriction of the map f to the set $\mathcal{X}'$ of possible values of X (the set of all $x \in \mathcal{X}$ such that $\mathbf{P}(X = x) > 0$) is injective.
e) Let Y be a discrete random variable on $\mathcal{Y}$ such that $H(Y \mid X) = 0$. Prove that there exists a function $g\colon \mathcal{X} \to \mathcal{Y}$ such that $\mathbf{P}(Y = g(X)) = 1$. (One says that the equality $Y = g(X)$ holds almost surely.)

Observe that the existence of such a function is equivalent to the existence, for every $x \in \mathcal{X}'$, of a unique $y \in \mathcal{Y}$ such that $q(x, y) > 0$, where q is the law of the joint variable (X, Y).

1.5. Entropy rate of Markov processes. 51

Exercise 1.6 (Shuffling increases entropy). [p. 154]   Let X be a discrete random variable with values in a pack of cards, identified with the set $\mathcal{X} = \{1, \ldots, 52\}$. We also consider a discrete random variable S with values in the group $\mathfrak{S}_{52}$ of permutations of $\mathcal{X}$. We assume that S and X are independent and we set $Y = S(X)$. The discrete random variable Y thus represents a pack of cards randomly shuffled.

 a) Assume that the random variable S is uniform. Prove that Y is uniform; what is its entropy? Explain why $H(Y) \geq H(X)$.

 b) Prove that the inequality $H(Y) \geq H(X)$ holds in general. (*Justify the equality* $H(X, S) = H(Y, S)$, *and evaluate both sides.*)

Exercise 1.7 (Pick and throw, I). [p. 155]   We consider a bag of two coins, that we denote by a and b. Each of them, unfair, has some probability of showing heads, say p and q, where $0 < p, q < 1$. We pick at random (uniformly) one coin from the bag, and toss it many times. Let Y be the random variable with values in $\{a, b\}$ that indicates the chosen coin, and let X_n be the random variable with values in {heads, tails} that indicates the result of the n-th coin toss.

 a) By computing their probability laws, prove that X_1 and X_2 have the same law.

 b) Does it hold that the variables X_1 and X_2 are independent? Formulate your answer in terms of p and q.

 c) Compute $I(X_1, X_2 \mid Y)$ and $I(X_1, X_2)$.

Exercise 1.8 (Sums of random variables). [p. 156]   Let X and Y be discrete random variables with values in **R** and let $Z = X + Y$.
 a) Prove that $H(Z) \leq H(X) + H(Y)$.
 b) Prove that $H(Z \mid X) = H(Y \mid X)$.
 c) Assuming that X and Y are independent, prove that $\sup\bigl(H(X), H(Y)\bigr) \leq H(Z)$.
 d) Give an example where $\inf\bigl(H(X), H(Y)\bigr) > H(Z)$.

Exercise 1.9. [p. 157]   Given two discrete random variables X and Y, following the same law, and assumed to have nonzero finite entropy, we set $\rho(X; Y) = 1 - H(Y \mid X)/H(X)$.

 a) Prove that $\rho(X; Y) = I(X, Y)/H(X)$. Conclude that ρ is symmetric.

 b) Prove that $\rho(X; Y) \in [0, 1]$. What does it mean if $\rho(X; Y) = 0$? 1?

Exercise 1.10. [p. 158]   For any two discrete random variables X, Y with finite entropy, we set $\rho(X; Y) = H(X|Y) + H(Y|X)$.
 a) Check that $\rho(X, Y) = H(X, Y) - I(X, Y) = 2H(X, Y) - H(X) - H(Y)$.
 b) Prove that ρ satisfies the following properties, where X, Y, Z are discrete random variables with finite entropy.
 (i) $\rho(X, Y) \geq 0$;
 (ii) $\rho(X, Y) = \rho(Y, X)$;
 (iii) $\rho(X, Z) \leq \rho(X, Y) + \rho(Y, Z)$.

Under what condition do we have $\rho(X, Z) = \rho(X, Y) + \rho(Y, Z)$?

 c) Find a necessary and sufficient condition on X, Y so that $\rho(X, Y) = 0$.

The map ρ satisfies the properties that we usually ask for a distance, with the important exception that it may vanish on pairs of distinct random variables.

Exercise 1.11. [p. 158]   Let $X = (X_n)_{n \in \mathbf{N}}$ be a stochastic process on a finite set A of cardinality $a \geq 1$.

a) Prove that the upper entropy rate $\overline{H}(X)$ is bounded from above by $\log(a)$.

b) Is this bound optimal?

Exercise 1.12 (Pick and throw, II). [p. 159]   We go back to the notation and context of exercise 1.7. Let $X = (X_n)_{n \in \mathbf{N}^*}$ be the random process consisting of the successive tosses of the initially chosen coin.

a) Compute $\lim_{n \to +\infty} \frac{1}{n} H(X_1, \ldots, X_n \mid Y)$.

b) What is the entropy rate $H(X)$?

c) Is X a homogeneous Markov chain?

Exercise 1.13 (Convergence to a stationary law). [p. 161]   Let us consider the homogeneous 2-state Markov chain studied in example 1.5.5, page 43. Let us denote by $\mu_n = (u_n, v_n)$ the law of the random variable X_n. If the law μ_0 is stationary, then the sequence (μ_n) is constant. We consider here the general case and study the convergence of the sequence (μ_n).

a) Compute μ_n in terms of the transition matrix P and of the initial law μ_0.

b) Does the sequence $(\mu_n)_{n \in \mathbf{N}}$ admit a limit when $(p,q) = (1,1)$ or $(p,q) = (0,0)$? Compute this limit if it exists.

From now on, we assume that $0 < p + q < 2$.

c) Compute the eigenvalues of P.

d) Find a basis of $\mathbf{R}^2$ consisting of eigenvectors for P, then compute P^n for every integer n.

e) Deduce the limit of μ_n when $n \to +\infty$. What do you observe?

Exercise 1.14. [p. 162]   We consider an independent sequence (X_n) of random variables that follow the Bernoulli law with parameter p; for $n \geq 1$, we set $Y_n = X_n + X_{n-1}$.

a) Compute the law of Y_n and its entropy.

b) Compute the conditional entropy $H(Y_n \mid Y_{n-1})$.

c) Is the random process (Y_n) markovian?

d) What is the entropy rate of the random process (Y_n)? (*Condition to X_0.*)

Exercise 1.15. [p. 164]   a) Let X, X', Y, Y' be random variables with values in the same finite set A. Prove that

$$D((X, X'), (Y, Y')) = D(X, Y) + \sum_{a} P(X = a) D((X' \mid X = a), (Y' \mid Y = a)).$$

(The latter divergence is that of the law of Y' conditioned to the event $(Y = a)$ with respect to the law of X' conditioned to the same event.) In particular, prove that $D((X, X'), (Y, Y')) \geq D(X, Y)$.

b) Let $(X_n)_{n \geq 0}$ and $(Y_n)_{n \geq 0}$ be homogeneous Markov chains with set of states A, and having the same transition matrix P. We respectively denote by p_n and q_n the laws of X_n and Y_n.

1.5. Entropy rate of Markov processes.

c) Prove that the sequence $(D(p_n, q_n))_{n \geq 0}$ decreases.

d) Describe this sequence when the given Markov chains are primitive, assuming moreover that (Y_n) is stationary.

e) We assume moreover that the law of p_n is the uniform law on A. Prove that the sequence $(H(X_n))$ of entropies increases.

Exercise 1.16. [p. 165] We consider two discrete random variables T and X, with values in sets Θ and A respectively. We assume that we know the conditional law of X given T, that is, that we know the probabilities $P(X = a \mid T = t)$, and we wish to estimate the value of T from that of X. This is an important problem for statistics, where T would correspond to some quantity of interest that we cannot measure directly, and X to some possible measurements.

Let f be a function on A. One says that f is a *sufficient statistic* for T if one has $I(T, f(X)) = I(T, X)$.

a) Prove that the inequality $I(T, f(X)) \leq I(T, X)$ holds in general, and that f is a sufficient statistic for T if and only if X and T are conditionally independent with respect to $f(X)$.

b) We assume that $\Theta = [0; 1]$, $X = (X_1, \ldots, X_n)$ and that for any t, conditionally to the event $(T = t)$, the X_k are independent Bernoulli random variables with parameter t.

Prove that the function f defined by $f(a_1, \ldots, a_n) = a_1 + \cdots + a_n$ is a sufficient statistic for T.

c) Let $\theta \colon A \to \Theta$ be any function; we consider that $\theta(X)$ is an estimator for T. If f is a sufficient statistic for T, we denote by $\theta^* \colon A \to \Theta$ the map $a \mapsto E(\theta(X) \mid f(X) = f(a))$. Using the results of exercise 0.7, page 22, prove the relations

$$E(\theta^*(X)) = E(\theta(X))$$

and

$$V(\theta^*(X) - T) = V(\theta(X) - T) - E(V(\theta(X) \mid f(X))).$$

In particular, this establishes the *Rao–Blackwell inequality*:

$$E((\theta^*(X) - T)^2) \leq E((\theta(X) - T)^2).$$

Exercise 1.17. [p. 167] This is about a student who reads a book in a library and, with probability p, has a coffee; then, before she returns to her studies, she may, with probability q, go outside for a few minutes for some fresh air.

a) Describe a 3-state Markov chain that models this story; draw the quiver that represents it. You may denote the states by b, c, a.

b) Give its transition matrix; check that it is a stochastic matrix.

c) Prove that this Markov chain possesses a unique stationary law and determine it.

d) Assume that X_0 follows this stationary law; what is then the entropy rate of the associated stochastic process?

e) We assume that $0 < p < 1$ and $0 < q < 1$, and that X_0 is certain with value b. What is the entropy rate of the associated stochastic process?

Exercise 1.18. [p. 168]   Chess is played on an 8×8 grid, and the king can move from its position to any of the immediately neighboring squares; it thus has eight possibilities inside of the chessboard, five on the edges, and three on the corners. We imagine an erratic king who, every second, moves in a random way to a possible square, with equal probabilities, each move being independent of the preceding ones. Let (X_n) be the random process where X_n represents the position of the king at time n.

a) Prove that (X_n) is a primitive Markov chain. (*Consider the quiver whose vertices are the squares of the chessboard, and where edges link a square to each of its neighbors.*)

b) Compute the stationary probability law. (*Look for a stationary probability law whose value only depends on the type of square (interior, boundary, corner).*)

c) Compute its entropy rate.

Chapter 2.
Coding

We now study the two fundamental contributions of SHANNON (1948) to information theory by studying coding, that is, the search for optimal ways to transmit a message.

We first try to minimize the broadcasting time, which makes us study the question of file *compression*: how do we code them so that they fill as little space as possible, and can we make precise the point to which they can be compressed? This question had already been raised at the beginning of electrical transmission: the issue was to have several telegraphic signals use the same electric line. This is fundamental for our digital era: the EPUB format for texts, the FLAC, OGGVORBIS or MP3 formats for sounds, the JPEG or DJVU format for images all try to make use of the type of stored information to do it efficiently. However, we will not be studying the question of the easiness or speed of coding and decoding, although it is equally important for practical applications, especially for real time ones.

We then present a crucial theorem of SHANNON (1948) which he coined as the *approximate equirepartition property*: for certain stochastic processes, everything happens as if the random variables were uniformly chosen among a subset of the set of possible values. This interpretation is also at the heart of Shannon's proofs in his paper; however, his arguments lack rigor here, because he doesn't really quantify the meaning of "as if" in this instance.

Another aspect of coding studied in SHANNON (1948) is to take into account the possibility of interferences while broadcasting a signal, hence of transmission errors, due to the physical character of a transmission channel. We present the model of a memoryless channel, where the successive errors are considered as mutually independent, following a law that uniquely depends on the transmitted symbols. This model excludes, for example, saturation phenomena.

The *transmission capacity* of such a channel is defined using entropy, more precisely, the mutual information that the emitted signal and the received signal may share. In a memoryless channel, the possibility of an error cannot be absolutely avoided, but Shannon's remarkable theorem is that it can be made as small as possible, as soon as one does not hope to transmit more information than the transmission capacity of the channel. For the proof of this theorem, we move away from Shannon's arguments and make use of those of WOLFOWITZ (1958), which are more precise. A common feature of both proofs, however, is that they do not seek to explicitly build a coding/decoding process, but to show that a randomly chosen coding process works! This is the birth, simultaneously to the works of Erdős,[1] of the *probabilistic method* in mathematics. The fecundity of this method is witnessed by the book of ALON & SPENCER (2008).

2.1. Codes

2.1.1. Alphabets and words — Let A be a set. We view the elements of A as the letters of an *alphabet*, and we consider the words that we can write with these symbols. By definition, a *word* is a finite sequence of elements of A, an object of the form $(a_1, \ldots, a_n)$, where $a_1, \ldots, a_n$ belong to A; the integer n is the *length* of this word. We denote by A^* the set of all words written in the alphabet A; it is the union of the sets A^n of words of length n as n ranges over all integers. We denote by $\ell(a)$ the length of a word a.

There is a unique word of length 0, the empty word, sometimes denoted by ε.

The set A^* is endowed with an internal law, concatenation of words. Given words $a = (a_1, \ldots, a_n)$ and $b = (b_1, \ldots, b_m)$, the word ab is given by the sequence $(a_1, \ldots, a_n, b_1, \ldots, b_m)$. Its length is the sum of the lengths of the words a and b. This law is associative; the empty word is its neutral element.

2.1.2. — A *code* C on the alphabet A, with values in an alphabet B, is a map from A to B^* such that $\ell(C(a)) > 0$ for all a. By such a code, the symbols from A are represented by nonempty words in the alphabet B.

A significant example consists in taking for alphabet A a vast enough set of symbols, containing, say, all letters from the Latin script and punctuation symbols, for alphabet B the set $\{0, 1\}$, and for code C the map which associates with a symbol of A its code in the ASCII system, or in one of the Unicode systems, written in binary notation.

Codes with values in the 2-element set $\{0, 1\}$ are also called *binary codes*.

2.1.3. — Let C be a code on the alphabet A. In practice, it is not enough to just code symbols from the alphabet A, but we have to code words in that alphabet. We do this by concatenating the codes of the symbols that constitute a word: if $(a_1, \ldots, a_n)$ is a word, its code is the concatenation $C(a_1) \ldots C(a_n)$. We denote by C^*, or sometimes even C, the map from A^* to B^* so defined. It satisfies $C^*(ab) = C^*(a)C^*(b)$ for all words a, b in A^*.

---

[1] Paul ERDŐS (1913–1996) was a Hungarian mathematician, whose prolific mathematical work encompassed number theory, graph theory and combinatorics. Being a Jew, he had to exile himself to Great Britain in 1934, where he then started a nomadic life, passing from one institute to another, according to the invitations he received. We also owe to him the idea of a Book in which God has collected the most beautiful proofs. The book (AIGNER & ZIEGLER, 2018) has been written as a homage: readers who like short, striking, elegant mathematical results will certainly read it with delight.

2.1. Codes.   57

One says that the code C is *uniquely decodable* if this map C^* is injective, that is, if two distinct words have distinct codes.

2.1.4. Prefix codes — We say that a code C on an alphabet A is a *prefix code* if for any two letters $a, b \in A$ such that $a \neq b$, neither of the words $C(a)$ or $C(b)$ is at the head of the other one.

Let us prove that such a word is uniquely decodable. Indeed, consider words $a = (a_1, \ldots, a_n)$ and $b = (b_1, \ldots, b_m)$ in the alphabet A such that $C^*(a) = C^*(b)$, and let us prove that $a = b$. By assumption, we have $C(a_1) \ldots C(a_n) = C(b_1) \ldots C(b_m)$.

If $a = \varepsilon$, then $C(a) = \varepsilon$, so that the words $C(b_1), \ldots, C(b_m)$ are empty. This can only happen if $m = 0$ since, by definition of a code, the code of a symbol is not empty. Therefore, $a = \varepsilon = b$. Similarly, if $b = \varepsilon$, then $a = \varepsilon$.

Let us now assume that $n \geqslant 1$; by what precedes, we also have $m \geqslant 1$. Let $p = \ell(C(a_1))$; up to exchanging the roles of a and b, we may assume that $p \leqslant \ell(C(b_1))$. By definition of the words $C(a)$ and $C(b)$, the word of B^* consisting of the first p symbols of $C(a)$ is equal to $C(a_1)$. Since $C(a_1)C(a_2) \ldots C(a_n) = C(a) = C(b) = C(b_1) \ldots C(b_m)$ and $\ell(C(b_1)) \geqslant p$, it is also the word consisting of the first p symbols of $C(b_1)$, so that the word $C(b_1)$ starts with $C(a_1)$. Since C is a prefix code, we have $a_1 = b_1$. The words obtained from $C^*(a)$ and $C^*(b)$ by removing the first p symbols are then equal too, which means that $C(a_2) \ldots C(a_n) = C(b_2) \ldots C(b_m)$. By induction, we have $n - 1 = m - 1$, hence $n = m$, and $a_2 = b_2, \ldots, a_n = b_n$.

We thus have proved that $a = b$.

Example 2.1.5. — Let us assume that $A = \{a, b, c\}$ and $B = \{0, 1\}$. The code C defined by $C(a) = 00$, $C(b) = 01$ and $C(c) = 1$ is a prefix code, since none of the words 00, 01, or 1 is the head of another one. Let us explain how to decode the word 0010001, that is, how to find a word m in the alphabet A such that $C^*(m) = 0010001$. If this holds, then m is not empty, since the empty word is only coded by an empty word; let then x be the first symbol of m. By definition of the map C^*, the word 0010001 starts with $C(x)$; the only possible symbol is $x = a$, so that m starts with a. Let us then write $m = am'$, for some word m'. Since $C^*(m) = 0010001 = C^*(am') = C(a)C^*(m') = 00C^*(m')$, we have $C^*(m') = 10001$. As before, the word m' is nonempty, and it has to start with c; writing $m' = cm''$, we get $C^*(m'') = 0001$. By the same reasoning, we obtain $m'' = ab$, hence finally $m = acab$.

This argument also allows us to prove that a word of B^* does not belong to the image of C^*. Let us, for example, look for a word m such that $C^*(m) = 100010$. This word has to start with cab, and if we write it as $m = cabm'$, for some word $m' \in A^*$, then we will have $C^*(m') = 0$. This word m' cannot be empty; it thus starts with one of the words $C(a), C(b), C(c)$, which is absurd.

Remark 2.1.6. — One also says that a prefix code is *instantaneously decodable*, by which it is meant that to start decoding a word, it is not necessarily to have received it in full: for a word $m \in A^*$, to determine the first symbol of m, we only need to recognize, at the head of $C^*(m)$, one of the words $C(x)$ for $x \in a$. Then we write $m = xm'$, for some word $m' \in A^*$, and it remains to decode the word $C^*(m')$ that we obtain by removing the word $C(x)$ from the head of $C^*(m)$.

2.2. The Kraft–McMillan inequality

Proposition 2.2.1 (Kraft[2], McMillan[3]). — *Let A be a set, let C be a code on A with values in a finite alphabet B, and set D = Card(B). If the code C is uniquely decodable, then the following inequality holds:*

$$\sum_{a \in A} D^{-\ell(C(a))} \leq 1.$$

Proof. — To prove this inequality, we may assume that the set A is finite. Let N be an integer such that $N \geq \ell(C(a))$ for every $a \in A$.

Let k be an integer such that $k \geq 1$. We have

$$\left(\sum_{a \in A} D^{-\ell(C(a))}\right)^k = \sum_{(a_1,\ldots,a_k) \in A^k} D^{-\ell(C(a_1))} \ldots D^{-\ell(C(a_k))}$$
$$= \sum_{a \in A^k} D^{-\ell(C(a))}.$$

For every integer m, let us denote by c_m the number of words $a \in A^k$ of length k such that $C(a)$ has length $\leq m$. By assumption, the map from A^k to B^* given by $(a_1, \ldots, a_k) \mapsto C(a_1) \ldots C(a_k)$ is injective. Consequently, c_m is at most the number of words of B^* of length $\leq m$, so that $c_m \leq 1 + D + \cdots + D^m = (D^{m+1} - 1)/(D - 1) \leq D^{m+1}$. We also have $c_m = 0$ for $m > kN$. Therefore,

$$\sum_{a \in A^k} D^{-\ell(C(a))} = \sum_m c_m D^{-m} \leq \sum_{m=1}^{kN} D^{m+1} D^{-m} = kND,$$

from which it follows that

$$\sum_{a \in A} D^{-\ell(C(a))} \leq (kND)^{1/k}.$$

When we make k tend to $+\infty$, we obtain the desired inequality. □

Theorem 2.2.2 (Shannon). — *Let X be a discrete random variable with values in a set A. Let C be a code on an alphabet A, with values in a finite alphabet B with cardinality $D \geq 2$. If C is uniquely decodable, then the average length of C(X) satisfies the inequality*

---

[2] I couldn't obtain biographical information regarding Leon KRAFT beyond his M. Sc. Thesis (MIT, 1949, http://hdl.handle.net/1721.1/12390), in which he proves this inequality, as well as its converse using an analysis he attributes to R. REDHEFFER.

[3] Brockway MCMILLAN (1915–2016) was an American scientist and public servant. He was hired by Bell Labs in 1946 and proved this inequality in a 1956 paper. Besides his work in mathematics, he also headed the Society for Industrial and Applied Mathematics (SIAM) as well as the American military research effort.

2.2. The Kraft–McMillan inequality.

$$\mathbf{E}\bigl(\ell(C(X))\bigr) \geqslant H_D(X),$$

where $H_D(X)$ is the base D entropy of X. Equality holds if and only if $\ell(C(a)) = -\log_D(\mathbf{P}(X = a))$ for every $a \in A$ such that $\mathbf{P}(X = a) > 0$.

Proof. — We shall deduce Shannon's inequality from that of Kraft–McMillan. By definition of the average length of $C(X)$ and of the entropy of X, we have

$$\mathbf{E}\bigl(\ell(C(X))\bigr) - H_D(X) = \sum_{a \in A} \mathbf{P}(X = a)\,\ell(C(a)) + \sum_{a \in A} \mathbf{P}(X = a) \log_D(\mathbf{P}(X = a))$$

$$= -\sum_{a \in A} \mathbf{P}(X = a) \log_D\left(\frac{D^{-\ell(C(a))}}{\mathbf{P}(X = a)}\right).$$

Since the logarithm function is concave, we have

$$\sum_{a \in A} \mathbf{P}(X = a) \log_D\left(\frac{D^{-\ell(C(a))}}{\mathbf{P}(X = a)}\right) \leqslant \log_D\left(\sum_{a \in A} \mathbf{P}(X = a) \frac{D^{-\ell(C(a))}}{\mathbf{P}(X = a)}\right)$$

$$= \log_D\left(\sum_{a \in A} D^{-\ell(C(a))}\right) \leqslant 0,$$

since, by the Kraft--McMillan inequality, the argument of the logarithm is $\leqslant 1$. Consequently, $\mathbf{E}\bigl(\ell(C(X))\bigr) - H_D(X) \geqslant 0$, hence Shannon's inequality. The logarithm function is strictly concave and strictly increasing, so that equality holds if and only if it holds at each step of the reasoning: on the one hand, this gives $\sum D^{-\ell(C(a))} = 1$, and on the other hand, all terms $D^{-\ell(C(a))}/\mathbf{P}(X = a)$, for $a \in A$ such that $\mathbf{P}(X = a) > 0$, have to be equal. Since $\sum \mathbf{P}(X = a) = 1$, this means $D^{-\ell(C(a))} = \mathbf{P}(X = a)$, for every a such that $\mathbf{P}(X = a) > 0$, hence $\ell(C(a)) = -\log_D(\mathbf{P}(X = a))$ for every such a. □

2.2.3. Efficient codes — Let A be an alphabet, let p be a discrete probability law on A. Let B be a finite alphabet with cardinality $D \geqslant 2$.

By Shannon's inequality, we have $\mathbf{E}\bigl(\ell(C(X))\bigr) \geqslant H_D(X)$, for every random variable X with values in A: entropy gives an unbreakable limit to the compression of a message. We now want to show that this limit can almost be reached, moreover by a prefix code! We first start with a converse to the Kraft–McMillan inequality.

Proposition 2.2.4. — *Let A be a set and let D be an integer such that $D \geqslant 2$. We consider a map $\ell \colon A \to \mathbf{N}^*$ for which the inequality*

$$\sum_{a \in A} D^{-\ell(a)} \leqslant 1$$

holds. Then there exists a prefix code C on A with values in an alphabet of cardinality D such that $\ell(C(a)) = \ell(a)$ for every $a \in A$.

Proof. — Let us enumerate the elements of the set A as a sequence $a_1, a_2, \ldots$ so that $\ell(a_1) \leqslant \ell(a_2) \leqslant \ldots$ This is obviously possible when A is finite. When A is infinite, we observe that for any n, there are finitely many $a \in A$ such that $\ell(a) = n$, for otherwise the sum $\sum_{a \in A} D^{-\ell(a)}$ would be infinite; it then suffices to first enumerate the elements a such that $\ell(a) = 1$, then those for which $\ell(a) = 2$, etc.

Let us define a strictly increasing sequence of rational numbers by setting

$$z_n = \sum_{m < n} D^{-\ell(a_m)},$$

for any integer n such that $n \leqslant \mathrm{Card}(A)$. Since $\sum_{a \in A} D^{-\ell(a)} \leqslant 1$, we have $z_n \leqslant 1 - D^{-\ell(a_n)} < 1$ for any integer $n \leqslant \mathrm{Card}(A)$. Consider the base D expansion of z_n: it takes the form $z_n = 0, y_1 y_2 \ldots y_p$, where the integer p satisfies $p \leqslant \ell(a_{n-1})$, and $y_1, \ldots, y_p$ are integers such that $0 \leqslant y_j \leqslant D - 1$ for all j, and $y_p \neq 0$. Let us associate with the symbol a_n the code $C(a_n) = y_1 \ldots y_p 0 \ldots 0$ in the alphabet $\{0; 1; \ldots; D-1\}$, ending with $\ell(a_n) - p$ symbols 0 so that $\ell(C(a_n)) = \ell(a_n)$.

Let m, n be integers such that $m < n \leqslant \mathrm{Card}(A)$. We have

$$z_n - z_m = \sum_{q=m}^{n-1} D^{-\ell(a_q)} \geqslant D^{-\ell(a_m)}.$$

Consequently, the base D expansions of z_m and z_n differ no later than by their $\ell(a_m)$-th base D digits, so that $C(a_m)$ is not a prefix of $C(a_n)$. On the other hand, $C(a_n)$ is not a prefix of $C(a_m)$: this is obvious if $\ell(a_n) > \ell(a_m)$, and when $\ell(a_m) = \ell(a_n)$, this would imply $C(a_m) = C(a_n)$ which cannot hold, as we just saw.

The map $C \colon A \to \{0; \ldots; D-1\}^*$ that we have defined is thus a prefix code, and one has $\ell(C(a)) = \ell(a)$ for all a. □

Example 2.2.5. — Assume that $A = \{a, b, c, d, e\}$ and suppose that we have $\ell(a) = \ell(b) = 1, \ell(c) = 2, \ell(d) = \ell(e) = 3$; we enumerate A as $a_1 = a, a_2 = b, \ldots, a_5 = e$. Take $D = 3$. The base 3 expansions of the rational numbers $z_1, \ldots, z_5$ introduced in the proof of the proposition write $z_1 = 0$, $z_2 = 0.1$, $z_3 = 0.2$, $z_4 = 0.21$ and $z_5 = 0.211$. We thus can set $C(a) = 0$, $C(b) = 1$, $C(c) = 20$, $C(d) = 210$ and $C(e) = 211$.

Theorem 2.2.6 (Shannon). — *Let X be a discrete random variable with values into a set A; we assume that $P(X = a) > 0$ for all $a \in A$. Let B be a finite set of cardinality $D \geqslant 2$. There exists a prefix code C on A, with values in B, such that*

$$H_D(X) \leqslant E\bigl(\ell(C(X))\bigr) < H_D(X) + 1.$$

In order that there exists such a code such that $E\bigl(\ell(C(X))\bigr) = H_D(X)$, it is necessary and sufficient that for every $a \in A$, the probability $P(X = a)$ be of the form D^{-m}, for some integer $m \geqslant 0$; we then have $\ell(C(a)) = -\log_D\bigl(P(X = a)\bigr)$.

Proof. — For every $a \in A$, let $\lambda(a) = \lceil -\log_D (P(X = a)) \rceil$, the smallest integer which is greater than or equal to $\log_D (P(X = a))$; in particular, we have $D^{-\lambda(a)} \leq P(X = a)$. Then,
$$\sum_{a \in A} D^{-\lambda(a)} \leq \sum_{a \in A} P(X = a) = 1.$$
By proposition 2.2.4, there exists a prefix code C on A, with values in
$$\{0; 1; \ldots; D-1\},$$
such that $\ell(C(a)) = \lambda(a)$ for all $a \in A$. Let us prove that this code satisfies the conclusion of the theorem.

The inequality $H_D(X) \leq E(\ell(C(X)))$ is a particular case of theorem 2.2.2. On the other hand, we have
$$\begin{aligned} E(\ell(C(X))) &= \sum_{a \in A} P(X = a)\ell(C(a)) \\ &= \sum_{a \in A} P(X = a)\lceil \log_D (P(X = a)) \rceil \\ &< \sum_{a \in A} P(X = a)\left(\log_D (P(X = a)) + 1\right) \\ &= H_D(X) + \sum_{a \in A} P(X = a) = H_D(X) + 1. \end{aligned}$$

This concludes the proof.

The last assertion follows from this, and from theorem 2.2.2: indeed, equality holds in Shannon's inequality if and only if $\ell(C(a)) = -\log_D (P(X = a))$ for all $a \in A$. Moreover, if $P(X = a)$ is of the form D^{-m}, we have $\lambda(a) = -\log_D (P(X = a))$ and the code that we have defined satisfies $E(\ell(C(X))) = H_D(X)$. □

2.3. Optimal codes

2.3.1. — Although its average length is close to the limit imposed by entropy, the prefix code constructed in the proof of Shannon's theorem 2.2.6 is not necessarily optimal. For example, if X follows a Bernoulli law with parameter $p \in \,]0; 1[$ and the goal alphabet has two symbols, so that $D = 2$, then the code words have lengths $\lceil -\log_2(p) \rceil$ and $\lceil -\log_2(1-p) \rceil$, so that their average length is equal to $S(p) = p\lceil -\log_2(p) \rceil + (1-p)\lceil -\log_2(1-p) \rceil$. On the other hand, we could just make a code with the two obvious words of length 1, getting a code of average length 1, which would be better, even strictly better unless $p = 1/2$.

More precisely, assume that $p < 1/2$ (the other case is symmetric) and consider the unique integer $m \geq 1$ such that $2^{-m-1} \leq p < 2^{-m}$, then $\lceil -\log_2(p) \rceil = m + 1$, while $\lceil -\log_2(1-p) \rceil = 1$ because $1 - p > 1/2$. Then $S(p) = p(m+1) + (1-p) =$

$mp + 1 < 1 + m/2^m < 1.5$. And when $p \to 1/2$, while remaining strictly smaller to $1/2$, the average length $S(p)$ tends to 1.5. On the other hand, the longest length of a code word is equal to $m + 1$, and becomes arbitrarily large when p tends to 0. However, the probability of that event tends to 0 and this compensates so that the average length of a code word tends to 1 in that case.

Definition 2.3.2. — *Let C be a uniquely decodable code on an alphabet A with values in an alphabet B. Let p be a discrete probability law on A such that $p(a) > 0$ for every $a \in A$. One says that the code C is* optimal *(with respect to p and B) if C minimizes the expression $\sum_{a \in A} p(a)\ell(C(a))$ among all uniquely decodable codes on A with values in the alphabet B.*

The expression $\ell_p(C) = \sum_{a \in A} p(a)\ell(C(a))$ is the average length of the code of a symbol from A, when these symbols are randomly chosen to follow the law p.

Lemma 2.3.3. — *Let A be a finite alphabet and let p be a probability law on A such that $p(a) > 0$ for every $a \in A$. Let B be a finite alphabet.*

a) *There exists a prefix code on A with values in B which is optimal.*

b) *If C is an optimal code (with respect to p and B), and if a, b are elements of A such that $p(a) < p(b)$, then $\ell(C(a)) \geq \ell(C(b))$: less probable symbols have longer codes.*

c) *If C is a prefix code which is optimal, then for every symbol $a \in A$ such that $C(a)$ has maximal length, there exists another symbol $b \in A$ such that $C(b)$ has the same length as $C(a)$ and differs only in the last symbol.*

Proof. — a) The elements $a \in A$ such that $p(a) = 0$ do not intervene in the definition of the average length of a code; up to replacing A by its subset consisting of elements a such that $p(a) > 0$, we assume that $p(a) > 0$ for every $a \in A$.

Let C_1 be a uniquely decodable code (for example, one given by Shannon's theorem) and let C be a second code which is uniquely decodable and for which $\ell_p(C) \leq \ell_p(C_1)$. In particular, we have the inequality $p(a)\ell(C(a)) < \ell_p(C_1)$, for every $a \in A$, hence $\ell(C(a)) < \ell_p(C_1)/p(a)$. This proves that the code C takes its values in the finite set of words of length $\leq \ell_p(C_1)/\inf(p)$. The set of all these maps is finite, so there are only finitely many uniquely decodable codes whose average length is smaller than or equal to that of C_1. In this finite set, we may find one of minimal average length, and it is optimal.

Let C be an optimal code. Its lengths $\ell(C(a))$ satisfy the Kraft–McMillan inequality $\sum D^{-\ell(C(a))} \leq 1$. Consequently, there exists a *prefix* code C' such that $\ell(C'(a)) = \ell(C(a))$ for all $a \in A$. The code C' is uniquely decodable, and it has the same average length as C, hence it is optimal.

b) Let C be an optimal code, and let $a, b \in A$ be such that $p(a) < p(b)$ and $\ell(C(a)) < \ell(C(b))$. Let us consider the code C' which coincides with C on $A - \{a, b\}$ but which exchanges the codes of a and b: $C'(a) = C(b)$ and $C'(b) = C(a)$. We then have

2.3. Optimal codes.

$$\begin{aligned}
\ell_p(C) - \ell_p(C') &= \sum_{x \in A} p(x)\ell(C(x)) - \sum_{x \in A} p(x)\ell(C'(x)) \\
&= p(a)\ell(C(a)) + p(b)\ell(C(b)) \\
&\quad - p(a)\ell(C'(a)) - p(b)\ell(C'(b)) \\
&= p(a)\ell(C(a)) + p(b)\ell(C(b)) \\
&\quad - p(a)\ell(C(b)) - p(b)\ell(C(a)) \\
&= (p(a) - p(b))\big(\ell(C(a)) - \ell(C(b))\big) \\
&> 0.
\end{aligned}$$

This contradicts the hypothesis that C is an optimal code.

c) Let C be an optimal code which is a prefix code. Let a be an element of A whose code word has maximal length; let us write $C(a) = mx$, where $m \in B^*$ and $x \in B$ (so that x is the last symbol of $C(a)$). Assume, by contradiction, that m is not the head of any other code word and let us then consider the code C' which coincides with C on $A - \{a\}$ and such that $C'(a) = m$. It is still a prefix code: by assumption, $C'(a)$ isn't the prefix of any other code word, and another code word, say $C'(b) = C(b)$, cannot be a prefix of $C'(a) = m$ since it would then be a prefix of $C(a) = mx$. In particular, the code C' is uniquely decodable, but its average length is strictly smaller than the average length of C, contradicting the hypothesis that C is optimal. Consequently, m is the prefix of another code word, say $C(b)$; let us write $C(b) = mp$, where $p \in B^*$. Since $C(a) = mx$ has maximal length, equal to $\ell(m) + 1$, we have

$$\ell(p) = \ell(C(b)) - \ell(m) \leq \ell(C(a)) - \ell(m) = 1,$$

so that either p is the empty word, or p consists of a single symbol. If p is empty, then $C(b) = m$ is a prefix of $C(a) = mx$, contradicting the hypothesis that C is a prefix code. Consequently, there exists a $y \in B$ such that $p = (y)$, and the words $C(a) = mx$ and $C(b) = my$ only differ by their last symbol. □

2.3.4. Huffman coding — Let A be a finite set and let $(p(a))_{a \in A}$ be a discrete probability law on A such that $p(a) > 0$ for every $a \in A$. A *Huffman code*[4] (with respect to the law p) is a binary code which is defined as follows by induction on the cardinality of A. (We say "a" Huffman code because the following construction involves several choices which we leave unspecified.)

If $\mathrm{Card}(A) = 1$, there is only one symbol to code, and we choose whatever code word of length 1 we want, either 0 or 1. If $\mathrm{Card}(A) = 2$, the code H associates with the two symbols of A the words 0 and 1, both of length 1. Let us now assume that

---

[4] David HUFFMAN (1925–1999) was an American computer scientist. In 1951, Prof. R. FANO suggested to his students, in place of the usual final exam, to think about optimal codes—without telling them he hadn't himself succeeded in solving the question. Huffman codes were the answer of his student D. HUFFMAN! He then wrote a PhD Thesis in electric engineering. He made an academic career at the computer science departments of MIT and of the University of California (Santa Cruz).

Card(A) > 2 and let a, b be two elements of A which minimize p: explicitly, we have $p(a), p(b) \leq \inf_{c \neq a,b} p(c)$. Let A′ be the union of the set $A - \{a, b\}$ and of an auxiliary element, denoted by ab; we define a discrete probability law p' on A′ by $p'(c) = p(c)$ for $c \in A - \{a, b\}$, and $p(ab) = p(a) + p(b)$. Let H′ be a Huffman code (for the law p' on the alphabet A′). The code H_p associates with a symbol $c \in A - \{a, b\}$ the word H′(c); if $m = $ H′(ab) is the code of the symbol ab for H′, we set H$(a) = m0$ and H$(b) = m1$.

Example 2.3.5. — Let us assume that $A = \{a, b, c, d, e\}$, with probabilities given by the table

a	b	c	d	e
0.25	0.25	0.20	0.15	0.15

Huffman's method starts by combining d and e and associating with them the probability $p'(de) = 0.30$, the other symbols remaining a, b, c, with their initial probabilities. This gives the table:

a	b	c	de
0.25	0.25	0.20	0.30

Then it combines the symbols a and c (here we have some choice), and associates with them the probability $p''(ac) = 0.45$, the other symbols being b and de, with probabilities 0.25 and 0.30:

ac	b	de
0.45	0.25	0.30

Finally, it combines b and de, leading to the two symbols bde and ac, with probabilities 0.55 and 0.45.

ac	bde
0.45	0.55

We now go through this path in reverse. In the last step, we code ac by 0, and bde by 1. In the penultimate step, we code ac by 0, b by 10, and de by 11. Then we code c by 00, a by 01, b by 10, and de by 11. Finally, we obtain the code:

a	b	c	d	e
01	10	00	110	111

The base 2 entropy of a discrete random variable X that follows the law p is approximately equal to 2.285. Indeed, we have

$$H_2(X) = -2 \times 0.25 \log_2(0.25) - 0.20 \log_2(0.20) - 2 \times 0.15 \log_2(0.15) \approx 2.285.$$

2.3. Optimal codes.

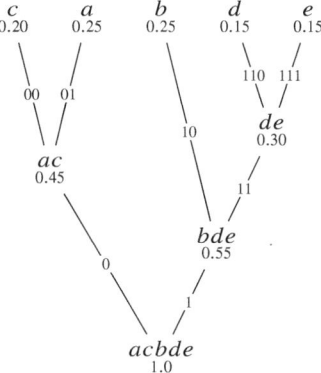

Fig. 2.1 Representation of the Huffman code by a tree

On the other hand, the average length of the above constructed Huffman code is equal to

$$\mathbf{E}\big(\ell_H(X)\big) = (2 \times 0.25 + 0.20) \times 2 + (2 \times 0.15) \times 3 = 2.3.$$

As a comparison, the lengths of the code words for the code constructed by Shannon's method are 2, 2, 3, 3, 3, and its average length is equal to

$$\mathbf{E}\big(\ell_S(X)\big) = (2 \times 0.25) \times 2 + (0.20 + 2 \times 0.15) \times 3 = 2.5.$$

Here is an example of such a code

a	b	c	d	e
00	01	100	101	110

Proposition 2.3.6 (D. Huffman, 1952). — *Let A be a finite set and let p be a discrete probability law on A. Every Huffman code (with respect to the law p on A) is a prefix code, and is optimal with respect to p.*

Proof. — Let H be a Huffman code with respect to p. Let us prove by induction on the cardinality of A that this code H is a prefix code. This is obvious if Card(A) $\leqslant$ 2, and we now assume that Card(A) $\geqslant$ 3. We make use of the notion introduced in the construction: a, b are two elements of A with minimal probabilities, the set A' is the union of A $-$ $\{a, b\}$ and of a symbol ab, and the probability law p' on a' associates with every $c \neq a, b$ the probability it had for p, and with the new symbol ab the sum of the probabilities of a and b. We then consider a Huffman code on A' for the probability law p', so that the code H is defined by $H(c) = H'(c)$ for $c \neq a, b$, $H(a) = H'(ab)0$ and $H(b) = H'(ab)1$.

By induction, the code H' is a prefix code. Still by induction, if c, c' are distinct elements of A, distinct from a and b, then $H(c) = H'(c)$ and $H(c') = H'(c')$ are not prefixes of one another. The words $H(a) = H'(ab)0$ and $H(b) = H'(ab)1$ are not prefixes of one another, because they are distinct and have the same length. For $c \neq a, b$, the word $H(c) = H'(c)$ is not a prefix of the word $H'(ab)$, by the induction hypothesis; if it were a prefix of any of the words $H'(ab)0$ or $H'(ab)1$, it would be equal to them, but then $H'(ab)$ would be a prefix of $H'(c)$, which contradicts the induction hypothesis that H' is a prefix code. Finally, if $H(a) = H'(ab)0$ or $H(b) = H'(ab)1$ is a prefix of $H(c)$, for some $c \neq a, b$, then $H'(ab)$ is a prefix of $H'(c) = H(c)$, yet another contradiction. This proves that H is a prefix code.

Let us now prove that the code H is optimal with respect to the law p. Let C be a uniquely decodable binary code which is optimal; by definition, its average length is smaller than that of H, which means that we have

$$\sum_{x \in A} p(x)\ell(C(x)) \leq \sum_{x \in A} p(x)\ell(H(x)).$$

Let us prove that equality holds, which will prove that H is optimal.

We may assume that C is a prefix code.

Let then $x \in A$ be an element such that $C(x)$ has maximal length. If we exchange $C(a)$ and $C(x)$, we obtain a new code C_1 whose average length satisfies

$$\ell_p(C_1) - \ell_p(C) = p(a)\ell(C(x)) + p(x)\ell(C(a)) - p(a)\ell(C(a)) - p(x)\ell(C(x))$$
$$= (p(a) - p(x))(\ell(C(x)) - \ell(C(a)))$$
$$\leq 0,$$

since $p(a)$ is minimal. The average length of this new code is smaller than or equal to that of C, hence is equal because the code C was assumed to be optimal. It is still a prefix code. Replacing C by C_1, we may thus assume that $C(a)$ has maximal length.

Since the code C is an optimal prefix code, there exists a symbol $y \neq a$ such that $C(a)$ and $C(y)$ only differ by their last symbol (lemma 2.3.3, c)). If we exchange $C(b)$ and $C(y)$, we obtain a code C_2 whose average length satisfies

$$\ell_p(C_2) - \ell_p(C) = p(b)\ell(C(y)) + p(y)\ell(C(b)) - p(b)\ell(C(b)) - p(y)\ell(C(y))$$
$$= (p(b) - p(y))(\ell(C(y)) - \ell(C(b)))$$
$$\leq 0,$$

since $p(b) \leq p(y)$ and $\ell(C(y)) = \ell(C(a)) \geq \ell(C(b))$. Similarly, the average length of the code C_2 is less than or equal to that of C, hence is equal, since C is optimal. We may thus replace C by C_2 and assume that $C(b)$ only differs from $C(a)$ by its last symbol. Up to exchanging $C(a)$ and $C(b)$, which does not change the average length of C, we also assume that $C(a)$ ends by 0 and $C(b)$ ends by 1.

Our Huffman code H on A is defined via some Huffman code H' on the alphabet $A' = A - \{a, b\} \cup \{ab\}$, endowed with the probability law p'. $C'(x) = C(x)$ for $x \in A - \{a, b\}$, and by defining $C'(ab)$ as the word $C(a)$ deprived of its last symbol. This is a prefix code. Indeed, if x, y are distinct elements of $A - \{a, b\}$, we have

$C'(x) = C(x)$ and $C'(y) = C(y)$, so $C'(x)$ is not a prefix of $C'(y)$ since C is a prefix code. If $x \in A - \{a, b\}$, then $C'(x) = C(x)$ is not a prefix of $C'(ab)$, since otherwise, $C(x)$ would be a prefix of $C(a)$ because $C'(ab)$ i is a prefix of $C(a)$; since $C(a)$ is a code word of maximal length and $C'(ab)$ is a prefix of $C(a)$, while being not equal to $C(a)$, the word $C'(ab)$ is strictly smaller than $C'(x) = C(x)$, hence is not a prefix of it. The average length of the code C' (with respect to the law p') is given by

$$\ell_{p'}(C') = \sum_{x \neq a,b} p'(x)\ell(C'(x)) + p'(ab)\ell(C'(ab))$$
$$= \sum_{x \neq a,b} p(x)\ell(C(x)) + (p(a) + p(b))\big(\ell(C(a)) - 1\big)$$
$$= \ell_p(C) - (p(a) + p(b)),$$

because $\ell(C(a)) = \ell(C(b))$. The same computation for the Huffman code H' shows that the average length of H' with respect to the law p' is equal to

$$\ell_p(H) - (p(a) + p(b)).$$

By induction, the Huffman code H' is optimal with respect to the law p', so that $\ell_{p'}(C') \geq \ell_{p'}(H')$. We thus have $\ell_p(C) \geq \ell_p(H)$, hence equality, as was to be shown. □

2.4. The law of large numbers, and compression

In his original paper, SHANNON (1948) used an alternative description of entropy, related to the law of large numbers in the theory of probability. We start by recalling two important classical inequalities.

Proposition 2.4.1. — *Let X be a discrete random variable admitting an expectation.*

a) For every real number $t > 0$, we have the Markov *inequality:*

$$P(|X| > t) \leq E(|X|)/t.$$

b) Let us assume, moreover, that X possesses a variance $V(X)$. Then for every real number $t > 0$, we have the Bienaymé[5]–Chebyshev[6]*inequality:*

$$P(|X - E(X)| > t) \leq V(X)/t^2.$$

---

[5] Irénée-Jules BIENAYMÉ (1796–1878) was a French statistician. Successor of Laplace, translator of Chebyshev, he developed the calculus of probabilities and statistics, up to their applications in finance and social statistics. He was the first to state the Bienaymé–Chebyshev inequality.

[5] Pafnuty CHEBYSHEV (1821–1894) was a Russian mathematician. His works in probability theory make him one of the pioneers of modern probability theory. He also had an important impact in

Proof. — a) Let A be the set of all elements ω of the sample space Ω such that $|X(\omega)| > t$; we need to bound $\mathbf{P}(A)$ from above. Let us observe that, on A, the random variable $|X|/t$ is at least 1; on its complement, it is at least 0. We thus have the inequality $\mathbf{E}(|X|/t) \geq 1 \times \mathbf{P}(A) + 0 \times \mathbf{P}(\complement A)$. Since $\mathbf{E}(|X|/t) = \mathbf{E}(|X|)/t$, this implies the Markov inequality.

b) We now consider the random variable $Y = X - \mathbf{E}(X)$. Since X has a variance, Y^2 has an expectation, equal to $\mathbf{V}(X)$ by definition. Let us apply to Y^2 the Markov inequality: we get $\mathbf{P}(Y^2 > t^2) \leq \mathbf{E}(Y^2)/t^2 = \mathbf{V}(X)/t^2$. Since $Y^2 > t^2$ is equivalent to $|X - \mathbf{E}(X)| > t$, this proves the Bienaymé–Chebyshev inequality. □

It is in this form that we will use these inequalities, but it is interesting to deduce from them right away the *law of large numbers*.

Theorem 2.4.2 (Law of large numbers). — *Let (X_n) be a sequence of independent discrete random variables, assumed to follow the same law, and to have a finite expectation. For any $n \geq 1$, let us set $S_n = (X_1 + \cdots + X_n)/n$. Then for every real number $t > 0$, we have*

$$\mathbf{P}(|S_n - \mathbf{E}(X_1)| > t) \to 0$$

when n tends to $+\infty$.

In the language of probability theory, one says that (S_n) *converges in probability to* $\mathbf{E}(X_1)$. In 1929, Kolmogorov proved a stronger version of this result, the *strong law of large numbers*, which states, under the same hypotheses, that the set of $\omega \in \Omega$ such that $(S_n(\omega))$ does not converge to $\mathbf{E}(X_1)$ has null probability; see for example KOLMOGOROV (1956).

Proof. — Since all the X_n have the same law, they have the same expectation. Replacing X_n with $X_n - \mathbf{E}(X_1)$, we replace S_n with $S_n - \mathbf{E}(X_1)$. It thus suffices to prove that $\mathbf{P}(|S_n| > t)$ tends to 0 under the additional assumption that $\mathbf{E}(X_1) = 0$.

Let us start by proving a precise upper bound for this probability under the assumption that the X_n have finite variance. In this case, S_n admits a variance too, given by

$$\mathbf{V}(S_n) = \mathbf{E}(S_n^2) = \frac{1}{n^2}\mathbf{E}\big((X_1 + \cdots + X_n)^2\big) = \frac{1}{n^2}\sum_{i,j}\mathbf{E}(X_iX_j).$$

For $i \neq j$, the random variables X_i and X_j are independent, so that

$$\mathbf{E}(X_iX_j) = \mathbf{E}(X_i)\mathbf{E}(X_j) = 0.$$

For $i = j$, we have $\mathbf{E}(X_i^2) = \mathbf{V}(X_i) = \mathbf{V}(X_1)$ since the X_i have the same law, hence the same variance. Consequently, $\mathbf{V}(S_n) = \mathbf{V}(X_1)/n$. (More generally, the variance

---

number theory by establishing the first landmark results towards the prime number theorem. In particular, he proved "Bertrand's postulate", according to which there is, between any integer and twice that number, at least one prime number.

2.4. The law of large numbers, and compression.

of a sum of independent random variables is the sum of their variances.) Let us apply the Bienaymé–Chebyshev inequality: for every real number $t > 0$, we have

$$\mathbf{P}(|S_n| > t) \leq \mathbf{E}\big((S_n/t)^2\big) \leq \mathbf{V}(S_n)/t^2.$$

It follows that $\mathbf{P}(|S_n| > t) \leq \mathbf{V}(X_1)/nt^2$. In particular, $\mathbf{P}(|S_n| > t)$ tends to 0.

The general case is more difficult and is proved by a truncation method.

Let t be a strictly positive real number and let us prove that $\mathbf{P}(|S_n| > t)$ tends to 0. Fix a real number $\varepsilon > 0$ and let us choose $\delta > 0$ so that

$$16\delta \mathbf{E}(|X_1|) < \varepsilon t^2.$$

Let us then define new random variables X'_k and X''_k by $X'_k(\omega) = X_k(\omega)$ if $|X_k(\omega)| \leq \delta n$, and $X'_k(\omega) = 0$ otherwise; let us also set $X''_k = X_k - X'_k$. (The notation is slightly misleading, because the X'_k depend on n.) We will bound from above the probabilities $\mathbf{P}(|X'_1 + \cdots + X'_n| > nt/2)$ and $\mathbf{P}(|X''_1 + \cdots + X''_n| > nt/2)$. The sum of these probabilities will give us an upper bound for $\mathbf{P}(|X_1 + \cdots + X_n| > nt)$ which will be arbitrarily small provided δ is taken small enough and n large enough. The underlying idea of this method is that the X'_k are bounded (by δn), in particular they have a finite variance, so that we can apply to them the first case, while the X''_k will be rare because X''_k is zero when X_k is not too large.

First of all, the $|X'_j|$ follow the same law, are independent and bounded above by δn. Consequently, they have a variance, which can be estimated by

$$\mathbf{V}(X'_j) = \mathbf{V}(X'_1) \leq \mathbf{E}\big((X'_1)^2\big) \leq \delta \mathbf{E}(|X'_1|)n \leq \delta \mathbf{E}(|X_1|)n.$$

Consequently,

$$\mathbf{V}(X'_1 + \cdots + X'_n) = n\mathbf{V}(X'_1) \leq \delta \mathbf{E}(|X_1|)n^2.$$

When n tends to $+\infty$, observe that X'_1 tends to X_1 almost surely, and its absolute value is bounded above by $|X_1|$; by Lebesgue's dominated convergence theorem, $\mathbf{E}(X'_1)$ tends to $\mathbf{E}(X_1) = 0$. For n large enough, we thus have $|\mathbf{E}(X'_1)| \leq \delta$, so that

$$\mathbf{E}(X'_1 + \cdots + X'_n)^2 = n^2 \mathbf{E}(X'_1)^2 \leq \delta \mathbf{E}(|X_1|)n^2,$$

hence

$$\mathbf{E}\big((X'_1 + \cdots + X'_n)^2\big) = \mathbf{V}(X'_1 + \cdots + X'_n) + \mathbf{E}(X'_1 + \cdots + X'_n)^2 \leq 2\delta \mathbf{E}(|X_1|)n^2.$$

Applying the Markov inequality to the random variable $(X'_1 + \cdots + X'_n)^2$, we deduce

$$\mathbf{P}(|X'_1 + \cdots + X'_n| > nt/2) \leq 8\delta \mathbf{E}(|X_1|)t^{-2} \leq \varepsilon/2, \qquad (2.4.2.1)$$

for all large enough integers n.

On the other hand,

$$\mathbf{P}(|X''_1 + \cdots + X''_n| > nt/2) \leq \mathbf{P}(X''_1 + \cdots + X''_n \neq 0) \leq n\mathbf{P}(X''_1 \neq 0),$$

since the X_j'' follow the same law. By construction, either $X_1'' = 0$, or $|X_1''| > \delta n$, so that
$$\mathbf{P}(X_1'' \neq 0) = \mathbf{P}(|X_1''| > \delta n) \leqslant \mathbf{E}(|X_1''|/\delta n) = \mathbf{E}(|X_1''|)\delta^{-1}n^{-1},$$
by the Markov inequality. We thus obtain
$$\mathbf{P}(|X_1'' + \cdots + X_n''| > nt/2) \leqslant \mathbf{E}(|X_1''|)\delta^{-1}. \tag{2.4.2.2}$$

When n tends to $+\infty$, X_1'' tends almost surely to 0, and one has $|X_1''| \leqslant |X_1|$, so that $\mathbf{E}(|X_1''|)$ tends to 0, by Lebesgue's dominated convergence theorem. Consequently, inequality (2.4.2.2) implies that
$$\mathbf{P}(|X_1'' + \cdots + X_n''| > nt/2) < \varepsilon/2$$
for n large enough. Combining these two upper bounds, it follows that for n large enough, the event $(|X_1 + \cdots + X_n| > nt)$ has probability $< \varepsilon$, which concludes the proof. $\square$

Theorem 2.4.3 (Shannon). — *Let (X_n) be a sequence of discrete random variables, with values in a common finite set A, independent and following the same law p; set $c = \sum_{a,b \in A} p_a p_b (\log(p_a/p_b))^2$. For every integer $n \geqslant 1$, let us endow the set A^n with the product law. Let ε be a real number such that $\varepsilon > 0$, and let A_ε^n be the set of all $(a_1, \ldots, a_n) \in A^n$ such that*
$$e^{-n(H(X_1)+\varepsilon)} \leqslant \mathbf{P}(X_1 = a_1, \ldots, X_n = a_n) \leqslant e^{-n(H(X_1)-\varepsilon)}.$$
Then we have
$$\mathbf{P}(A_\varepsilon^n) > 1 - \frac{c}{2n\varepsilon^2},$$
and
$$\left(1 - \frac{c}{2n\varepsilon^2}\right) e^{n(H(X_1)-\varepsilon)} \leqslant \mathrm{Card}(A_\varepsilon^n) \leqslant e^{n(H(X_1)+\varepsilon)}.$$

Let us reformulate this statement. There, $1 - \mathbf{P}(A_\varepsilon^n)$ is the probability of the complement of A_ε^n, and it is bounded from above by $c/2n\varepsilon^2$; when n is large, it is arbitrarily small. In other words, when n is large, most of the tuples $(a_1, \ldots, a_n)$ have a probability which is close to $e^{-nH(X_1)}$, and there are approximately $e^{nH(X_1)}$ such tuples. Roughly speaking, when we perform a large number of independent throws of these random variables, everything happens as if we had performed a sequence of independent random throws, uniformly chosen among $e^{nH(X_1)}$ possible values. This is the *statistical interpretation* of entropy.

Proof. — Up to modifying the sample space by removing a set of null probability, then the set A to the set of possible values of X_1, we assume that $\mathbf{P}(X_1 = a) > 0$ for every $a \in A$. Let then $\varphi \colon A \to \mathbf{R}$ be the map defined by $\varphi(a) = -\log(\mathbf{P}(X_1 = a))$. Since the X_k are independent and follow the same law, we have

2.4. The law of large numbers, and compression.

$$P(X_1 = a_1, \ldots, X_n = a_n) = P(X_1 = a_1) \cdots P(X_n = a_n)$$
$$= P(X_1 = a_1) \cdots P(X_1 = a_n),$$

hence

$$-\frac{1}{n}\log\left(P(X_1 = a_1, \ldots, X_n = a_n)\right) = -\frac{1}{n}\sum_{k=1}^n \log\left(P(X_1 = a_k)\right) = \frac{1}{n}\sum_{k=1}^n \varphi(a_k).$$

Let us endow A with the law of X_1, and A^n with the product law: then $p(a_1, \ldots, a_n) = \prod_{k=1}^n P(X_1 = a_k)$ for all $a_1, \ldots, a_n$ in A. On this sample space A^n, we set $U_k(a_1, \ldots, a_n) = \varphi(a_k)$. The random variables $U_1, \ldots, U_n$ are independent, and they follow the same law. Indeed, for every $a \in A$, we have

$$P(U_k = \varphi(a)) = P(X_1 = a),$$

and $P(U_k = t) = 0$ if $t \notin \varphi(A)$. Since they take only finitely many values, they admit an expectation and a variance. Moreover, we have

$$H(X_1) = -\sum_{a \in A} P(X_1 = a) \log\left(P(X_1) = a\right) = E(U_1).$$

By the Bienaymé–Chebyshev inequality, we have

$$P\left(\left|H(X_1) - \frac{1}{n}\sum_{k=1}^n U_k\right| > \varepsilon\right) < V(U_1)/n\varepsilon^2.$$

We have

$$\frac{1}{n}\sum_{k=1}^n U_k(a_1, \ldots, a_n) = -\frac{1}{n}\log\left(P(X_1 = a_1, X_2 = a_2, \ldots, X_n = a_n)\right).$$

It remains to estimate the variance $V(U_1)$ of U_1. By definition, we have

$$V(U_1) = E\left((U_1 - E(U_1))^2\right) = E(U_1^2) - E(U_1)^2,$$

so that, writing $p_a = P(X_1 = a)$, we have

$$V(U_1) = \sum_{a \in A} p_a \log(p_a)^2 - \sum_{a,b \in A} p_a p_b \log(p_a) \log(p_b)$$
$$= \sum_{a,b \in A} p_a p_b \left(\log(p_a)^2 - \log(p_a)\log(p_b)\right)$$
$$= \sum_{a,b \in A} p_a p_b \log(p_a) \log(p_a/p_b).$$

By symmetry, we also have

$$\mathbf{V}(U_1) = \sum_{a,b \in A} p_a p_b \log(p_b) \log(p_b/p_a).$$

Adding these two relations, we obtain the equality

$$2\mathbf{V}(U_1) = \sum_{a,b \in A} p_a p_b \log(p_b/p_a)^2 = c.$$

This concludes the proof of the first inequality.

Let us then use the defining upper bound of $\mathbf{P}(X_1 = a_1, \ldots, X_n = a_n)$, for $(a_1, \ldots, a_n) \in A_\varepsilon^n$: this implies

$$1 \geqslant \mathbf{P}(A_\varepsilon^n) \geqslant \text{Card}(A_\varepsilon^n) e^{-n(H(X_1)+\varepsilon)},$$

hence the asserted upper bound for $\text{Card}(A_\varepsilon^n)$. Using the analogous lower bound, we also get

$$1 - \frac{c}{2n\varepsilon^2} \leqslant \mathbf{P}(A_\varepsilon^n) \leqslant \text{Card}(A_\varepsilon^n) e^{-n(H(X_1)-\varepsilon)},$$

which implies the asserted lower bound for $\text{Card}(A_\varepsilon^n)$. □

2.4.4. — How did Shannon deduce from this theorem the possibility of compressing a signal (within the entropy limit)? Let us fix a parameter $\varepsilon > 0$ and let us consider the "typical set" A_ε^n of A^n defined in theorem 2.4.3. Its cardinality is at most $2^{n(H_2(X_1)+\varepsilon)}$; so that, by enumerating its elements and writing these numbers in binary notation, each of them only requires $\lceil n(H_2(X_1)+\varepsilon) \rceil$ bits. The other elements may require $\text{Card}(A)^n$ symbols, hence $n \log_2(\text{Card}(A))$ bits, but they only appear with a small probability, smaller than $c/2n\varepsilon^2$. Consequently, the average length of a code of a sequence of n symbols is bounded above by

$$\lceil n(H_2(X_1) + \varepsilon) \rceil + \frac{c}{2n\varepsilon^2} n \log_2(\text{Card}(A)).$$

Once divided by n, we obtain

$$\frac{1}{n}\lceil n(H_2(X_1) + \varepsilon) \rceil + \frac{c}{2n\varepsilon^2} \log_2(\text{Card}(A)),$$

a quantity smaller than $H_2(X_1) + 1$ when n is large.

2.4.5. — Similarly as for the entropy, the mutual information has a statistical interpretation. Let A and B be finite sets, and let $C = A \times B$ be the product set. We consider independent random values

$$Z_1 = (X_1, Y_1), \ldots, Z_n = (X_n, Y_n),$$

with values in C, following the same law. Let then

$$X = (X_1, \ldots, X_n), \quad Y = (Y_1, \ldots, Y_n), \quad \text{and} \quad Z = (Z_1, \ldots, Z_n).$$

2.4. The law of large numbers, and compression.

Identifying the element $((a_1, b_1), \ldots, (a_n, b_n))$ of C^n with the element (a, b) of $A^n \times B^n$, we write $Z = (X, Y)$. Set

$$c_Z = \sum_{a,b \in C} \mathbf{P}(Z_1 = a)\mathbf{P}(Z_1 = b) \log\left(\mathbf{P}(Z_1 = a)/\mathbf{P}(Z_1 = b)\right)^2,$$

and define c_X and c_Y in a similar way.

Let ε be a real number such that $\varepsilon > 0$; we consider the subset C_ε^n of C^n consisting of pairs (a, b) such that

$$e^{-n(H(X_1)+\varepsilon)} \leqslant \mathbf{P}(X = a) \leqslant e^{-n(H(X_1)-\varepsilon)}$$

$$e^{-n(H(Y_1)+\varepsilon)} \leqslant \mathbf{P}(Y = b) \leqslant e^{-n(H(Y_1)-\varepsilon)}$$

and

$$e^{-n(H(X_1,Y_1)+\varepsilon)} \leqslant \mathbf{P}(X = a, Y = b) \leqslant e^{-n(H(X_1,Y_1)-\varepsilon)}.$$

Theorem 2.4.6. — *With this notation, we have*

$$\mathbf{P}(Z \in C_\varepsilon^n) \geqslant 1 - \frac{c_X + c_Y + c_Z}{2n\varepsilon^2}$$

and

$$\mathrm{Card}(C_\varepsilon^n) \leqslant e^{n(H(X,Y)+\varepsilon)}.$$

Moreover, if $X'_1, \ldots, X'_n$ on the one hand, and $Y'_1, \ldots, Y'_n$ on the other hand, are random values with the same law than $X_1, \ldots, X_n$ and $Y_1, \ldots, Y_n$, but independent, then

$$\left(1 - \frac{c_X + c_Y + c_Z}{2n\varepsilon^2}\right) e^{-n(I(X,Y)+3\varepsilon)} \leqslant \mathbf{P}((X', Y') \in C_\varepsilon^n) \leqslant e^{-n(I(X,Y)-3\varepsilon)}.$$

Proof. — We endow the set C^n with the law of Z. Let $\varphi_Z: C \to \mathbf{R}_+$ be the random variable defined by

$$\varphi_Z(c) = \begin{cases} -\log\left(\mathbf{P}(Z = c)\right), & \text{if } \mathbf{P}(Z = c) > 0, \\ 0, & \text{otherwise;} \end{cases}$$

let us define $\varphi_X: A \to \mathbf{R}_+$ and $\varphi_Y: B \to \mathbf{R}_+$ in the same way. As in the proof of theorem 2.4.3, we prove that

$$\mathbf{P}(\complement C_\varepsilon^n) < \frac{c_X}{2n\varepsilon^2} + \frac{c_Y}{2n\varepsilon^2} + \frac{c_Z}{2n\varepsilon^2},$$

hence the desired lower bound for $\mathbf{P}(C_\varepsilon^n)$. We also prove, using the same argument, that

$$\left(1 - \frac{c_X + c_Y + c_Z}{2n\varepsilon^2}\right) e^{n(H(X_1,Y_1)-\varepsilon)} \leqslant \mathrm{Card}(C_\varepsilon^n) \leqslant e^{n(H(X_1,Y_1)+\varepsilon)}.$$

Let us now consider random variables X', Y', independent and following the same laws as X, Y, and let us set $Z' = (X', Y')$. We have

$$\begin{aligned} P(Z' \in C_\varepsilon^n) &= \sum_{(a,b) \in C_\varepsilon^n} P(Z' = (a,b)) \\ &= \sum_{(a,b) \in C_\varepsilon^n} P(X' = a) P(Y' = b) \\ &= \sum_{(a,b) \in C_\varepsilon^n} P(X = a) P(Y = b). \end{aligned}$$

By definition of C_ε^n and the upper bound for $\mathrm{Card}(C_\varepsilon^n)$, we thus have

$$\begin{aligned} P(Z' \in C_\varepsilon^n) &\leqslant \mathrm{Card}(C_\varepsilon^n) e^{-n(H(X_1)-\varepsilon)} e^{-n(H(Y_1)-\varepsilon)} \\ &\leqslant e^{-n(H(X_1)+H(Y_1)-H(X_1,Y_1)-3\varepsilon)} = e^{-n(I(X_1,Y_1)-3\varepsilon)}. \end{aligned}$$

The proof of the lower bound is analogous:

$$\begin{aligned} P(Z' \in C_\varepsilon^n) &\geqslant \mathrm{Card}(C_\varepsilon^n) e^{-n(H(X_1)+\varepsilon)} e^{-n(H(Y_1)+\varepsilon)} \\ &\geqslant \left(1 - \frac{c_X + c_Y + c_Z}{2n\varepsilon^2}\right) e^{-n(H(X_1)+H(Y_1)-H(X_1,Y_1)+3\varepsilon)} \\ &= \left(1 - \frac{c_X + c_Y + c_Z}{2n\varepsilon^2}\right) e^{-n(I(X_1,Y_1)+3\varepsilon)}. \end{aligned}$$

This concludes the proof of the theorem. □

2.5. Transmission capacity of a channel

The two themes that we have studied up to now, entropy of a random variable and coding, only concerned the first two steps of the communication diagram that we presented in the introduction. We now consider the third one, the transmission channel. In particular, we wish to take into account the problem of *noise*, which can cause a received signal to differ from the emitted one.

This noise is out of our control, but we have to *model* it, that is, to give it a mathematical description. The probabilistic model postulates that it is a random phenomenon.

Definition 2.5.1. — *Let A and B be alphabets. A memoryless transmission channel from alphabet A to alphabet B is given by a family $(p(\cdot \mid a))$ of probability laws on B, indexed by the set A.*

For $a \in A$ and $b \in B$, the number $p(b \mid a)$ is the probability that the channel transmits the symbol b given that the emitted symbol was a. The matrix $(p(b \mid a))$ of type $A \times B$ is the matrix of *transmission probabilities* of the memoryless channel;

2.5. Transmission capacity of a channel.

its rows, indexed by the elements a of A, give the laws $p(\cdot \mid a)$. In particular, its coefficients are positive, and the sum of each row is equal to 1.

We describe this channel as *memoryless* because for this model of noise, the transmission of a word $(a_1, \ldots, a_n)$ is done symbol by symbol, in a random and independent way. In other words, the probability that the word $(a_1, \ldots, a_n)$ will be transmitted as $(b_1, \ldots, b_n)$ is given by

$$\mathbf{P}(Y = b_1 \ldots b_n \mid X = a_1 \ldots a_n) = p(b_1 \mid a_1) \cdots p(b_n \mid a_n).$$

In particular, if the emitter submits a random word $X = (X_1, \ldots, X_n)$ to the channel and the receiver obtains $Y = (Y_1, \ldots, Y_n)$, the random variable Y_n is conditionally independent of the variables $X_1, \ldots, X_{n-1}$ and $Y_1, \ldots, Y_{n-1}$, with respect to X_n. More generally, for every subset $\{i_1, \ldots, i_p\}$ of $\{1, \ldots, n\}$: the variable $(Y_{i_1}, \ldots, Y_{i_p})$ is conditionally independent of the variables X_j and Y_j (for $j \neq i_1, \ldots, i_p$) with respect to $(X_{i_1}, \ldots, X_{i_p})$.

Definition 2.5.2. — *Let C be a memoryless channel with matrix of transmission probabilities $(p(b \mid a))$. We define the* transmission capacity *of this channel to be the expression*

$$I(C) = \sup I(X, Y),$$

where X runs among all random variables with values in A, and Y runs among all random variables with values in B, such that for all $a \in A$ and $b \in B$, we have $\mathbf{P}(Y = b \mid X = a) = p(b \mid a)$.

In what follows, if C is a memoryless transmission channel from an alphabet A to an alphabet B, with transition probabilities $p(b \mid a)$, we will denote by $X \sim_C Y$ to mean that X and Y are random variables with values in A and B respectively such that, for every $a \in A$ and every $b \in B$, we have $\mathbf{P}(Y = b \mid X = a) = p(b \mid a)$. We will also use this notation when X and Y take their values in A^n and B^n respectively, and then we mean that

$$\mathbf{P}(Y = b_1 \ldots, b_n \mid X = a_1 \ldots a_n) = \prod_{j=1}^{n} p(b_j \mid a_j).$$

For every pair (X, Y) of random variables with values in A and B respectively, we have the inequality

$$0 \leqslant I(X, Y) \leqslant \max\left(\log\left(\mathrm{Card}(A)\right), \log\left(\mathrm{Card}(B)\right)\right),$$

so that the transmission capacity of the channel C satisfies the inequality

$$0 \leqslant I(C) \leqslant \max\left(\log\left(\mathrm{Card}(A)\right), \log\left(\mathrm{Card}(B)\right)\right).$$

Let us finally recall the following formulas that relate entropy, conditional entropy and mutual information:

$$I(X, Y) = H(X) + H(Y) - H(X, Y)$$
$$= H(Y) - H(Y \mid X)$$
$$= H(X) - H(X \mid Y).$$

2.5.3. — We first explain the definition of the transmission capacity of a memoryless channel by considering the case, as did SHANNON (1948, §12), where we just have to transmit two symbols 0 and 1 *"at a rate of 1000 symbols per second with probabilities $p_0 = p_1 = \frac{1}{2}$."* Shannon goes on: *"Thus, our source is producing information at the rate of 1000 bits per second. During transmission the noise introduces errors so that, on the average, 1 in 100 is received incorrectly (a 0 as 1, or 1 as 0). What is the rate of transmission of information? Certainly less than 1000 bits per second since about 1% of the received symbols are incorrect. Our first impulse might be to say the rate is 990 bits per second, merely subtracting the expected number of errors. This is not satisfactory since it fails to take into account the recipient's lack of knowledge of where the errors occur. We may carry it to an extreme case and suppose the noise so great that the received symbols are entirely independent of the transmitted symbols. The probability of receiving 1 is $\frac{1}{2}$ whatever was transmitted and similarly for 0. Then about half of the received symbols are correct due to chance alone, and we would be giving the system credit for transmitting 500 bits per second while actually no information is being transmitted at all."*

He then explains how to measure the transmission capacity: *"the proper correction to apply to the amount of information transmitted is the amount of this information which is missing in the received signal, or alternatively the uncertainty when we have received a signal of what was actually sent. From our previous discussion of entropy as a measure of uncertainty it seems reasonable to use the conditional entropy of the message, knowing the received signal, as a measure of this missing information."*

This is precisely what the third equality of the preceding formula says: we obtain the mutual information $I(X, Y)$ by starting from the information contained in the sent message X, as measured by its entropy $H(X)$, and subtracting the uncertainty $H(X \mid Y)$ measured by the conditional entropy of X given Y. The second equality presents this quantity as the information $H(Y)$ of the received message deprived of the noise measured by the conditional entropy $H(Y \mid X)$.

Taking the supremum over all possible laws on symbols X reflects the possibility, for the source, of *adapting* the way in which he transforms his message into a signal to take into account the channel issue. For example, imagine if all symbols but one are transmitted without error, then it would be a good idea to consider a coding that does not make use of the faulty symbol.

Let us now give some examples of transmission channels, and how their transmission capacities can be computed.

Examples 2.5.4. — 1. Let us take $A = B = \{0, 1\}$ as alphabets and let p be a real number in the interval $[0; 1]$. The *symmetric binary transmission channel* with parameter p transmits an incorrect symbol with probability p. When $p = 0$, it transmits the emitted symbol faithfully: the channel is noiseless. When $p = 1/2$, this channel transmits any symbol with probability $1/2$, independently of the

2.5. Transmission capacity of a channel.

emitted one; we can guess that such a channel is not very useful, and we will see how its transmission capacity reflects it. The case $p = 1$ may look puzzling: the channel systematically modifies the emitted symbol, sending a "negative" version of the initial signal; there should be no loss of information in this case. The matrix of transmission probabilities of this channel is given by

$$\begin{pmatrix} 1-p & p \\ p & 1-p \end{pmatrix}.$$

Let X and Y be random variables on A and B respectively such that $X \sim_C Y$. Recall that this means that $\mathbf{P}(Y = b \mid X = a) = p(b \mid a)$ for all $a \in A$ and $b \in B$. Set $u = \mathbf{P}(X = 0)$, so that $\mathbf{P}(X = 1) = 1 - u$ and

$$H(Y \mid X) = uH(Y \mid X = 0) + (1 - u)H(Y \mid X = 1)$$
$$= uh(p) + (1 - u)h(1 - p) = h(p).$$

On the other hand, since Y is a binary random variable, we have $H(Y) \leq \log(2)$. Consequently,

$$I(X, Y) \leq \log(2) - h(p).$$

When X follows a uniform law ($u = 1/2$), so does Y—we can argue by symmetry, or make the computation, so that the transmission capacity of this channel is equal to $\log(2) - h(p)$. At this point, it is sensible to consider base 2 logarithms, so that entropies are measured in *bits*: then the transmission capacity of the symmetric binary transmission channel is equal to $I(C) = 1 - h(p)$.

When $p = 0$, we get $I(C) = 1$; when $p = 1/2$, we have $I(C) = 1/2$; and when $p = 1$, we again find $I(C) = 1$, in coherence with our discussion that in this case, the channel does not lose information.

2. A variant of the preceding channel uses the alphabet $A = \{0, 1\}$ but has for target alphabet the set $B = \{0, 1, e\}$, where e is an auxiliary symbol meant to indicate the system detected a faulty transmission, which happens with probability q, the probability of transmitting an incorrect symbol being $(1 - q)p$. In other words, we have $p(1 \mid 0) = p(0 \mid 0) = (1 - q)q$ and $p(e \mid 1) = p(e \mid 0) = q$. The matrix of transmission probabilities for this channel is thus given by

$$\begin{pmatrix} (1-q)(1-p) & (1-q)p & q \\ (1-q)p & (1-q)(1-p) & q \end{pmatrix}.$$

As in the preceding example, let us consider a pair (X, Y) of random variables such that $X \sim_C Y$; we evaluate $I(X, Y)$ by writing it as $H(Y) - H(Y \mid X)$. We start by evaluating the term $H(Y)$. Let us again set $\mathbf{P}(X = 0) = u$, and write $t = u(1 - p) + (1 - u)p$; we have

$$\mathbf{P}(Y = 0) = u(1 - q)(1 - p) + (1 - u)(1 - q)p = (1 - q)t,$$
$$\mathbf{P}(Y = 1) = (1 - u)(1 - q)(1 - p) + u(1 - q)p$$
$$= (1 - q)(1 - t),$$
$$\mathbf{P}(Y = e) = q.$$

Let E be the random variable with value 1 when $Y = e$ and 0 otherwise. Since E is a function of Y, we have $H(Y) = H(Y, E)$; then

$$H(Y) = H(Y, E) = H(E) + H(Y \mid E).$$

On the other hand, $\mathbf{P}(E = 1) = q$ and $\mathbf{P}(E = 0) = 1 - q$, so that $H(E) = h(q)$. Moreover, conditioned to the event $(Y = e)$, the random variable Y is certain, hence has zero entropy; conditioned to the complementary event, it follows a Bernoulli law with parameter t, so that

$$H(Y \mid E) = qH(Y \mid Y = e) + (1 - q)H(Y \mid Y \neq e) = (1 - q)h(t).$$

We thus have
$$H(Y) = h(q) + (1 - q)h(t).$$

We apply a similar argument to evaluate $H(Y \mid X)$: we have

$$H(Y \mid X) = H(Y, E \mid X) = H(E \mid X) + H(Y \mid E, X).$$

As above, the first term is equal to $h(q)$. Conditioned to the event $(Y = e)$, which has probability q, the random variable Y is certain; conditioned to the event $(E = 0) \cap (X = 0)$, which has probability $(1 - q)u$, it is a Bernoulli variable with parameter p; conditioned to the event $(E \neq 0) \cap (X = 0)$, which has probability $(1 - q)(1 - u)$, it also is a Bernoulli variable with parameter p. Consequently,

$$H(Y \mid E, X) = qH(Y \mid E = 1) + (1 - q)uH(Y \mid E = 0, X = 0)$$
$$+ (1 - q)(1 - u)H(Y \mid E = 0, X = 1)$$
$$= q \times 0 + (1 - q)u \times h(p) + (1 - q)(1 - u) \times h(p)$$
$$= (1 - q)h(p).$$

We then have
$$H(Y \mid X) = h(q) + (1 - q)h(p),$$

so that, finally,
$$I(X, Y) = (1 - q)\bigl(h(t) - h(p)\bigr).$$

The parameters p and q are fixed, hence this expression is maximal when $h(t)$ is maximal. Taking base 2 logarithms, we have $h(t) \leq 1$, so that $I(X, Y) \leq (1-q)\bigl(1-h(p)\bigr)$; we also have $h(t) = 1$ for $t = 1/2$. Since $t = u(1-p)+(1-u)p$, we see that $u = 1/2$ implies $t = 1/2$, so that

2.5. Transmission capacity of a channel.

$$I(C) = (1-q)(1-h(p)) \text{ bits.}$$

(Without this observation, we could simply have computed u in terms of t: we get $u = (t-p)/(1-2p)$, hence $u = 1/2$ for $t = 1/2$.)

3. We say that a channel is *weakly symmetric* if the rows of its matrix of transmission probabilities only differ from one another by permutations, and if the sum of each column is constant. It is said to be *symmetric* if, moreover, the columns of its matrix of transmission probabilities differ from one another by permutations; the symmetric binary channel of example 1 is an example of such a channel. However, unless $p = 1/3$, the channel of example 2 is not weakly symmetric.

One way to define a symmetric channel consists in taking $A = B = \mathbf{Z}/d\mathbf{Z}$ for alphabets (where $d \geqslant 2$ is an integer) and choosing transmission probabilities of the form $p(b \mid a) = q(b-a)$, where q is a given probability law on A. Of course, we may replace $\mathbf{Z}/d\mathbf{Z}$ by any finite group.

Let C be a weakly symmetric channel and let X, Y be random variables on A and B respectively, related by the condition $\mathbf{P}(Y = b \mid X = a) = p(b \mid a)$. For every $a \in A$, we have

$$H(Y \mid X = a) = -\sum_b p(b \mid a) \log\left(p(b \mid a)\right),$$

an expression which does not depend on a, thanks to the assumption on the rows of the matrix of transmission probabilities. On the other hand, we have

$$H(Y) \leqslant \log\left(\mathrm{Card}(B)\right).$$

When the law of X is uniform, the assumption on the sum of the coefficients of each column of the matrix of transmission probabilities implies that the law of Y is uniform too: indeed, for every $b \in B$, we have

$$\mathbf{P}(Y = b) = \sum_{a \in A} \mathbf{P}(Y = b \mid X = a)\mathbf{P}(X = a) = \frac{1}{\mathrm{Card}(A)} \sum_{a \in A} p(b \mid a),$$

and this expression does not depend on b. In this case, we thus have $H(Y) = \log\left(\mathrm{Card}(B)\right)$.

Given the relation $I(X, Y) = H(Y) - H(Y \mid X)$, we deduce that the transmission capacity of this channel is given by

$$I(C) = \log\left(\mathrm{Card}(B)\right) - H(Y \mid X = a),$$

where a is any element of A.

4. Let us consider a transmission channel C from an alphabet A to an alphabet B, and let n be an integer $\geqslant 2$; Let us define a memoryless transmission channel C^n from the alphabet A^n to the alphabet B^n by the transmission probabilities:

$$p(b \mid a) = \prod_{i=1}^{n} p(b_i \mid a_i),$$

for $a = (a_1, \ldots, a_n) \in A^n$ and $b = (b_1, \ldots, b_n) \in B^n$. Let us prove that $I(C^n) = nI(C)$.

Let $X = (X_1, \ldots, X_n)$ and $Y = (Y_1, \ldots, Y_n)$ be random variables with values in A^n and B^n respectively, and such that $\mathbf{P}(Y = b \mid X = a) = p(b \mid a)$ for $a \in A^n$ and $b \in B^n$. We start from the relation

$$I(X, Y) = H(Y) - H(Y \mid X),$$

and compute the first term by induction:

$$H(Y) = H(Y_1) + H(Y_2 \mid Y_1) + \cdots + H(Y_n \mid Y_1, \ldots, Y_{n-1})$$
$$\leqslant H(Y_1) + H(Y_2) + \cdots + H(Y_n),$$

since entropy decreases by conditioning; note that equality holds when the X_i are independent. Again by induction, we have

$$H(Y \mid X) = H(Y_1 \mid X) + H(Y_2 \mid Y_1, X) + \cdots + H(Y_n \mid Y_1, \ldots, Y_{n-1}, X).$$

Let $j \in \{1; \ldots; n\}$. By definition of the channel C^n, the random variable Y_j is conditionally independent of the Y_i and the X_i (for $i \neq j$) with respect to X_j, so that

$$H(Y_j \mid Y_1, \ldots, Y_{j-1}, X) = H(Y_j \mid X_j).$$

Therefore,

$$I(X, Y) \leqslant \sum_{j=1}^{n} H(Y_j) - \sum_{j=1}^{n} H(Y_j \mid X_j) = \sum_{j=1}^{n} I(X_j, Y_j),$$

with equality if the X_j are independent.

We thus obtain on the one hand an inequality $I(X, Y) \leqslant nI(C)$, so that $I(C^n) \leqslant nI(C)$, and on the other hand, assuming independence of the X_j chosen so that $I(X_j, Y_j) = I(C)$, the equality $I(X, Y) = nI(C)$. Finally, $I(C^n) = nI(C)$. If the supremum that defines $I(C)$ is not achieved by pairs (X_j, Y_j), we may always choose such pairs so that $I(X_j, Y_j)$ is arbitrarily close to $I(C)$, and conclude similarly.

2.6. Coding adapted to a transmission channel

2.6.1. — Unless its matrix of transmission probabilities is very specific (the precise condition is that for every $b \in B$, there exists at most one element $a \in A$ such that $p(b \mid a) > 0$), a transmission channel cannot transmit any message without a

2.6. Coding adapted to a transmission channel.

possibility of error. However, the theorem of SHANNON (1948) that we will now prove states that it is possible to arrange the signal so that the probability of an error is as small as desired, with a transmission speed only limited by the channel capacity.

Let us consider a transmission channel C from an alphabet A to an alphabet B; to fix our ideas, we may imagine that $A = B = \{0; 1\}$ and that C is the symmetric binary transmission channel that transmits an incorrect symbol with probability p.

We assume that the messages that we wish to send (a text, a video, a sound...) already have a digital structure: for example files in the MARKDOWN text format, video files in the H.264 format, music files in the FLAC format... The transmission process will essentially not depend on this internal structure.

With that aim, the source is willing to split its file into blocks of an appropriate size; each block will be *coded* as a word in the alphabet A which will be passed to the channel. The word transmitted to the receiver has to be *decoded*, and reconstructed one block after the other. As we shall see, the efficiency of the system relies on the fact that we may transmit words of a large enough length n.

Definition 2.6.2. — *Let C be a memoryless transmission channel from an alphabet A to an alphabet B. Let M be a finite set; a code Φ on M, of length n, for the transmission channel C is the datum of two maps $f_\Phi \colon M \to A^n$ and $g_\Phi \colon B^n \to M$.*

Again: the emitter rewrites its file as a series of words of length n in the alphabet A which are emitted through the channel; these words are of the form $\alpha = f_\Phi(m)$, where m is a "block"; the receiver obtains a word β of length n in the alphabet B and it decodes it using the function g_Φ, to get a block $g_\Phi(\beta)$. If the transmission had no error, one has $g_\Phi(\beta) = m$; otherwise, there was a transmission error, and the goal is to make its probability as small as possible.

The set M underlying this theoretical description has no importance at all: it only intervenes via the maps f_Φ and g_Φ, and each set with the same cardinality would be equally adequate.

The *transmission rate* of such a code is the quotient

$$\tau(\Phi) = \frac{\log\left(\text{Card}(M)\right)}{n}.$$

It is the amount of information that this code claims to transmit in relation to the number of symbols used.

The probability of a transmission error when one transmits a block $m \in M$ is given by

$$\lambda_m(\Phi) = \mathbf{P}\bigl(g_\Phi(Y) \neq m \mid X = f_\Phi(m)\bigr),$$

where X and Y are random variables with values in A^n and B^n respectively, related by the transmission probabilities defined by the channel C^n. By definition of a memoryless transmission channel, if $f_\Phi(m) = a_1 \ldots a_n$, we have

$$\lambda_m(\Phi) = \sum_{\substack{b \in B^n \\ g_\Phi(b) \neq m}} \mathbf{P}(Y = b \mid X = f_\Phi(m)),$$

$$= \sum_{\substack{b=(b_1,\ldots,b_n) \in B^n \\ g_\Phi(b) \neq m}} \prod_{i=1}^{n} p(b_i \mid a_i).$$

This shows that these probabilities of error only depend on the transmission probabilities of the channel C, and not on the choice of random variables X and Y which are adapted to the channel...

We also define the *maximal* probability of transmission error:

$$\lambda_{\max}(\Phi) = \sup_{m \in M} \lambda_m(\Phi),$$

and the *average* probability of transmission error:

$$\lambda_{\mathrm{av}}(\Phi) = \frac{1}{\mathrm{Card}(M)} \sum_{m \in M} \lambda_m(\Phi).$$

Definition 2.6.3. — *Let C be a memoryless transmission channel from an alphabet A to an alphabet B. One says that a real number ρ is an* accessible transmission rate *for the channel C if there exists, for every real number $\varepsilon > 0$ and every large enough integer n, a finite set M and a code Φ of length n on M with transmission rate $\geqslant \rho$ and maximal probability of transmission error $\leqslant \varepsilon$.*

The theorem of SHANNON (1948) relates the transmission capacity of a memoryless transmission channel (definition 2.5.2) and the accessible transmission rates for this channel.

Theorem 2.6.4 (Shannon). — *Let C be a memoryless transmission channel from an alphabet A to an alphabet B.*

Any accessible transmission rate for the channel C is less than or equal to the transmission capacity I(C) of this channel.

Conversely, any real number $\rho < \mathrm{I}(C)$ is an accessible transmission rate for this channel.

Proposition 2.6.5 (Fano[6] inequality). — *Let X, Z be discrete random variables with values in a finite set A. Set $\varepsilon = \mathbf{P}(X \neq Z)$; then*

$$\mathrm{H}(X \mid Z) \leqslant h(\varepsilon) + \varepsilon \log \big(\mathrm{Card}(A) - 1\big).$$

---

[6] Robert M. FANO (1917–2016) was an Italian-American MIT professor in electric engineering and computer science. His works encompassed information theory, distributed computing and microwaves.

2.6. Coding adapted to a transmission channel.

Proof. — Let U be the random variable such that $U = 1$ if $Z = X$ and $U = 0$ otherwise; it is a Bernoulli random variable with parameter ε, so that $H(U) = h(\varepsilon)$. By conditioning with respect to X, we have

$$H(X \mid Z) = H(X, U \mid Z) - H(U \mid X, Z).$$

Since U is certain given (X, Z), we have $H(U \mid X, Z) = 0$, so that

$$H(X \mid Z) = H(X, U \mid Z).$$

By conditioning with respect to U, we also have

$$H(X, U \mid Z) = H(U \mid Z) + H(X \mid U, Z).$$

Entropy decreases by conditioning, so that

$$H(U \mid Z) \leqslant H(U) = h(\varepsilon).$$

By definition of conditional entropy, we also have

$$H(X \mid U, Z) = (1 - \varepsilon) H(X \mid Z, U = 1) + \varepsilon H(X \mid Z, U = 0).$$

The first term vanishes, because if $U = 1$, the random variable X is certain given Z. For the second term, note that the entropy $H(X \mid Z, U = 0)$ is the entropy of a random variable with values in the set A, hence is at most equal to the entropy $\log(\operatorname{Card}(A))$ of a uniform random variable on A. More precisely, conditioned to the event $(U = 0)$, that is, $X \neq Z$, this random variable avoids one value, hence its entropy is at most $\log(\operatorname{Card}(A) - 1)$. Combining these inequalities, we thus have

$$\begin{aligned} H(X \mid Z) &= H(X, U \mid Z) \\ &= H(U \mid Z) + H(X \mid U, Z) \\ &\leqslant h(\varepsilon) + \varepsilon \log(\operatorname{Card}(A) - 1), \end{aligned}$$

hence the proposition. $\square$

2.6.6. — Let us first prove that any accessible transmission rate is at most $I(C)$. Let thus Φ be a code of length n on a finite set M for the channel C; let f and g be the coding and decoding functions for that code. Let W be uniform random variable with values in M; then $X = f(W)$ is a random variable with values in A^n that the channel transmits, and the word Y received on the other side is a random variable with values in B^n; the decoded symbol is then $W' = g(Y)$, which we need to compare with W. Set $\varepsilon = \mathbf{P}(W \neq W')$.

The random variable W is uniform in the set M, so that $H(W) = \log(\operatorname{Card}(M))$. Applied to the variables W, W', the Fano inequality (proposition 2.6.5) implies that $H(W \mid W') \leqslant h(\varepsilon) + \varepsilon \log(\operatorname{Card}(M))$. Using the relation

$$H(W) = H(W \mid W') + I(W, W'),$$

we thus obtain

$$\log\bigl(\mathrm{Card}(M)\bigr) \leqslant h(\varepsilon) + \varepsilon \log\bigl(\mathrm{Card}(M)\bigr) + \mathrm{I}(W, W'),$$

hence

$$(1 - \varepsilon) \log\bigl(\mathrm{Card}(M)\bigr) \leqslant h(\varepsilon) + \mathrm{I}(W, W').$$

In the chain of random variables $W \to X \to Y \to W'$, the random variables W and Y are conditionally independent given X (the channel does not know the word W with code $X = f(W)$), while the variables W and W' are conditionally independent given Y (because W' is certain given Y). By theorem 1.3.11, we thus have

$$\mathrm{I}(W, W') \leqslant \mathrm{I}(W, Y) = \mathrm{I}(Y, W) \leqslant \mathrm{I}(Y, X) = \mathrm{I}(X, Y).$$

Transmitting X to Y corresponds to the repeated channel C^n of example 2.5.4, so that its transmission capacity satisfies $\mathrm{I}(C^n) = n\mathrm{I}(C)$; then $\mathrm{I}(X, Y) \leqslant n\mathrm{I}(C)$, hence the inequality $\mathrm{I}(W, W') \leqslant n\mathrm{I}(C)$. We then have

$$(1 - \varepsilon) \log\bigl(\mathrm{Card}(M)\bigr) \leqslant h(\varepsilon) + n\mathrm{I}(C),$$

from which we deduce the inequality

$$\tau(\Phi) = \frac{\log\bigl(\mathrm{Card}(M)\bigr)}{n} \leqslant \frac{\mathrm{I}(C) + h(\varepsilon)/n}{1 - \varepsilon}.$$

Let us apply this inequality to codes of arbitrarily large length, (n tends to $+\infty$), and whose probability of transmission error is arbitrarily small (ε tends to 0, hence $h(\varepsilon)$ tends to 0); then the right-hand side tends to $\mathrm{I}(C)$, which implies that $\tau(\Phi) \leqslant \mathrm{I}(C)$: the transmission capacity of the channel C is an upper bound for its accessible transmission rates.

Remark 2.6.7. — WOLFOWITZ (1957) proved a more precise version of this part of the proposition: if the transmission rate of a code is strictly greater than its capacity, then the probability of a transmission error tends to 1, exponentially fast, when the length n of that code tends to infinity.

2.6.8. — Let us now prove the "positive" part of Shannon's theorem, namely, that every real number ρ such that $\rho < \mathrm{I}(C)$ is accessible for the channel C. We fix an integer $n \geqslant 1$ and a real number $\alpha > 0$; we need to prove that there exists, provided n is large enough, a code Φ of length n on a finite set M_φ of cardinality $\lceil \exp(n\rho) \rceil$ which is adapted to the channel C and whose maximal probability of transmission error is at most α. We shall first prove that there exists such a code whose average probability of transmission error is small, and we will then see how to deduce the existence of a code of the same length, whose maximal probability of transmission error is at most twice that of the original code, and whose transmission rate is only slightly smaller.

The method proposed by SHANNON (1948), and barely modified since, does not consist in constructing an explicit code adapted to the channel C, but it evaluates the

2.6. Coding adapted to a transmission channel.

expectation of the probability of a transmission error when the code Φ is chosen *at random*. Since this expectation is small, at least one of the envisioned codes has a small probability of transmission error.

More precisely: the set n will be large enough, M will be an unspecified set with cardinality $\lceil e^{n\rho} \rceil$, and what is random is the coding function $f_\Phi \colon M \to A^n$, chosen among all maps from M to A^n. For any $m \in M$, $f_\Phi(m)$ is a random variable on A^n; we assume that these random variables are independent, and that they all follow the law π on A^n, induced by a law on A that achieves the transmission capacity of the channel C. (Let us recall that I(C) is the upper bound, for all laws on A, of the mutual information I(X, Y), where X and Y are random variables with values in A and B respectively, related by the relation $\mathbf{P}(Y = b \mid X = a) = p(b \mid a)$, and $p(\cdot \mid \cdot)$ are the transmission probabilities of the channel C. Since A and B are finite, this upper bound is achieved by some law; this had not been the case, then we would have chosen a probability law which approximates it.)

The decoding function g_Φ is not random, that would be ridiculous, but follows the strategy of "typical correction". Although it is not explicit, the merit of this strategy is to allow for a quite simple evaluation of the probability of a transmission error.

In the set $C^n = A^n \times B^n$, let us consider the typical set C_ε^n with parameter $\varepsilon > 0$, as defined in theorem 2.4.3 regarding entropy, and in section 2.4.5 for mutual information. By definition, the decoding function g_Φ maps an element $b \in B^n$ to an element $m \in M$ such that $(f_\Phi(m), b) \in C_\varepsilon^n$, if there exists a unique such element, and to an unspecified element of M otherwise.

Let us recall the definition of the typical set: for $a \in A^n$ and $b \in B^n$, the pair (a, b) belongs to C_ε^n if and only if $\mathbf{P}(X = a, Y = b)$ is approximately equal to $e^{-nH(X,Y)}$, $\mathbf{P}(X = a)$ is approximately equal to $e^{-nH(X)}$, and $\mathbf{P}(Y = b)$ is approximately equal to $e^{-nH(Y)}$; the precise formulation of "is approximately equal to" depends on ε, but it is not yet necessary to make it explicit. The notation $\mathbf{P}(X = a, Y = b)$ is an abbreviation for the probability that the channel transmits the word b when given the word a:

$$\mathbf{P}(X = a, Y = b) = \mathbf{P}(X = a)\mathbf{P}(Y = b \mid X = a) = \prod_{j=1}^{n} \mathbf{P}(X = a_j) p(b_j \mid a_j).$$

The message word is also chosen randomly, by way of a uniform random variable W in M, independent of the code Φ. The word submitted to the channel is $X = f_\Phi(W)$, the word received is Y, and the decoded block is $W' = g_\Phi(Y)$.

To start with, we shall prove that the probability that $W' \neq W$ is small. Let U be the random variable equal to 1 when $W' \neq W$, and to 0 otherwise; we have $\mathbf{P}(W' \neq W) = \mathbf{E}(U)$.

Proposition 2.6.9. — *Let $\alpha > 0$. If $\varepsilon > 0$ is such that $\rho + 3\varepsilon < I(C)$, then for all large enough integers n, we have $\mathbf{P}(W' \neq W) \leqslant \alpha$.*

Proof. — Since $\mathbf{P}(W = m) = 1/\mathrm{Card}(M)$ for all $m \in M$, we have

$$P(W' \neq W) = \sum_{m \in M} P(W = m) P(W' \neq W \mid W = m)$$

$$= \frac{1}{\operatorname{Card}(M)} \sum_{m \in M} P(W' \neq W \mid W = m).$$

Let us fix $m \in M$ and look at the conditional probabilities given $W = m$; there is a transmission error when either $(f_\Phi(m), Y)$ does not belong to C_ε^n (event E; recall that $X = f_\Phi(W)$), or there exists $m' \neq m$ such that $(f_\Phi(m'), Y)$ belongs to C_ε^n (event $E_{m'}$), so that

$$P(W' \neq W \mid W = m) \leq P(E \mid W = m) + \sum_{m' \neq m} P(E_{m'} \mid W = m).$$

Then

$$P(W' \neq W) \leq P\big((X, Y) \notin C_\varepsilon^n\big)$$
$$+ \frac{1}{\operatorname{Card}(M)} \sum_{m \neq m'} P\big((f_\Phi(m'), Y) \in C_\varepsilon^n \mid X = f_\Phi(m)\big).$$

Let us consider a random variable X' with values in A^n, following the same law as X, but independent of X; by definition of a memoryless channel, the random variables X' and Y are independent. Let $m' \in M$ be such that $m' \neq m$; the random variable $f_\Phi(m')$ follows the same law as X and behaves as X', so that the pair $(f_\Phi(m'), Y)$ follows the same law as (X', Y). Therefore

$$\frac{1}{\operatorname{Card}(M)} \sum_{m \neq m'} P\big((f_\Phi(m'), Y) \in C_\varepsilon^n \mid X = f_\Phi(m)\big)$$
$$\leq \operatorname{Card}(M) P\big((X', Y) \in C_\varepsilon^n \mid X = f_\Phi(m)\big),$$

hence

$$P(W' \neq W) \leq P\big((X, Y) \notin C_\varepsilon^n\big) + P\big((X', Y) \in C_\varepsilon^n\big).$$

Let us define $c(X)$, $c(Y)$ and $c(X, Y)$ as in section 2.4.5; by the statistical interpretations of entropy (theorem 2.4.3) and of mutual information, the first term is bounded from above by $(c(X) + c(Y) + c(X, Y))/2n\varepsilon^2$, and the second one is bounded from above by

$$\operatorname{Card}(M) e^{-n(I(X_1, Y_1)) - 3\varepsilon} = \lceil e^{n\rho} \rceil e^{-n(I(C) - 3\varepsilon)} \sim e^{n(\rho - I(C) + 3\varepsilon)}.$$

By the choice of ε, we have $\rho - I(C) + 3\varepsilon < 0$. When n goes to $+\infty$, the upper bound that we established for $P(W' \neq W)$ tends to 0; for n large enough, we thus have $P(W' \neq W) < \alpha$. □

Corollary 2.6.10. — *Let C be a memoryless transmission channel and let $\alpha > 0$. For any real number $\rho < I(C)$ and any large enough integer n, there exists a code Φ*

2.6. Coding adapted to a transmission channel. 87

of length n which is adapted to the channel C whose transmission rate is at least ρ and whose average probability of transmission error $\lambda_{av}(\Phi)$ is less than α.

Proof. — Let us choose the integer n large enough so that $\mathbf{P}(W' \neq W) < \alpha$ (proposition 2.6.9). Conditioning on all possible coding functions, we have

$$\mathbf{P}(W' \neq W) = \frac{1}{\text{Card}(\{\Phi\})} \sum_{\Phi} \mathbf{P}(W' \neq W \mid f = f_{\Phi}).$$

Consequently, there exists an f_{Φ} such that $\mathbf{P}(W' \neq W \mid f = f_{\Phi}) < \alpha$. On the other hand, the random variable W being uniform in M, and independent on Φ, we have

$$\mathbf{P}(W' \neq W \mid f = f_{\Phi}) = \frac{1}{\text{Card}(M)} \sum_{m \in M} \mathbf{P}(W' \neq W \mid f = f_{\Phi}, W = m).$$

By definition, $\mathbf{P}(W' \neq W \mid f = f_{\Phi}, W = m) = \lambda_m(\Phi)$, the probability of transmission error when the block m is sent to the channel using the coding function f_{Φ}. Then

$$\lambda_{av}(\Phi) = \mathbf{P}(W' \neq W \mid f = f_{\Phi}) < \alpha.$$

Finally, the transmission rate of the code Φ satisfies

$$\tau(\Phi) = \frac{\log\bigl(\text{Card}(\Phi)\bigr)}{n} \geq \rho,$$

which concludes the proof of the corollary. □

Lemma 2.6.11. — *Let Φ be a code of length n adapted to a memoryless transmission channel C. There exists a code Φ' of the same length such that*

$$\tau(\Phi') \geq \tau(\Phi) - \frac{\log(2)}{n} \quad \text{and} \quad \lambda_{\max}(\Phi') \leq 2\lambda_{av}(\Phi).$$

Proof. — Let M be the domain of the code Φ, and let f and g be its coding and decoding functions.

Let M' be the set of all elements $m \in M$ whose probability of transmission error $\lambda_m(\Phi)$ satisfies $\lambda_m(\Phi) \leq 2\lambda_{av}(\Phi)$. Let us apply the Markov inequality (proposition 2.4.1) to the function $m \mapsto \lambda_m(\Phi)$ on the sample space M endowed with the uniform probability; its expectation is $\lambda_{av}(\Phi)$. Consequently,

$$\mathbf{P}\bigl(\lambda_m(\Phi) > 2\lambda_{av}(\Phi)\bigr) \leq \frac{1}{2},$$

that is, $\text{Card}(M - M') \leq \frac{1}{2}\text{Card}(M)$, hence $\text{Card}(M') \geq \frac{1}{2}\text{Card}(M)$.

Let $f': M' \to A^n$ be the restriction to M' of the coding function f. We choose any function $g': B^n \to M'$ such that $g'(b) = g(b)$ if $g(b) \in M'$. Then, (f', g') is a code Φ' on the set M' adapted to the channel C. For $m \in M'$, we have

$$\lambda_m(\Phi') = \mathbf{P}\big(g'(Y) \neq m \mid X = f'(m)\big)$$
$$\leqslant \mathbf{P}\big(g(Y) \neq m \mid X = f_\Phi(m)\big) = \lambda_m(\Phi) \leqslant 2\lambda_{\text{av}}(\Phi),$$

hence $\lambda_{\max}(\Phi') \leqslant 2\lambda_{\text{av}}(\Phi)$. Finally, the transmission rate of this code Φ' satisfies

$$\tau(\Phi') = \frac{\log\big(\text{Card}(M')\big)}{n} \geqslant \frac{\log\big(\text{Card}(M)\big) - \log(2)}{n} \geqslant \tau(\Phi) - \frac{\log(2)}{n}.$$

This concludes the proof. □

2.6.12. Conclusion of the proof of theorem 2.6.4 — Let us recall that we need to prove the existence, for any real number ρ such that $\rho < I(C)$, any real number $\alpha > 0$ and any large enough integer n, of a code of length n which is adapted to the channel C, with transmission rate at least ρ and maximal probability of transmission error at most α. Choose a real number ρ' such that $\rho < \rho' < I(C)$. By corollary 2.6.10, there exists, for any large enough integer n, a code Φ of length n which is adapted to the channel C, whose transmission rate is at least ρ' and average probability of transmission error at most $\alpha/2$.

Let Φ' be a code as given by lemma 2.6.11. Its transmission rate is at least $\rho' - \frac{\log(2)}{n}$; we thus have $\tau(\Phi') \geqslant \rho$ if n is large enough, specifically if $n \geqslant \log(2)/(\rho' - \rho)$. Its maximal probability of transmission error is at most $2 \times \alpha/2 = \alpha$. The theorem is then proved.

Remark 2.6.13. — The proof of Shannon's theorem relies on a probabilistic argument that implies the existence of efficient coding systems but does not provide the construction of any of them. Such constructions are the subject of the theory of *error correction codes*, initiated by the paper of HAMMING (1950). Exercise 2.14 explains the simplest of all Hamming[7] codes. The theory of error correction codes is of quite a different nature to the theory described in this book; linear algebra over finite fields plays an important role. An introduction to the field at the same level as the present book can be found in VANSTONE & OORSCHOT (1989).

However, the classical algebraic codes do not reach the efficiency allowed by Shannon's theorem. The law density parity codes (LDPC) invented by Gallager[8] reach this bound, but their actual implementation was not initially easy; since the 2000s, they have been used by several digital TV standards. In the 1990s, Claude Berrou[9] invented *turbocodes*; initially the first practical constructions to reach Shannon's bounds, they are widely used for 3G/4G and satellite telecommunication networks.

---

[7] Richard HAMMING (1915–1998) was an American mathematician whose works have been of great importance for the development of computer science, not only by the invention of error correction codes, but also by the study of the numerical integration of differential equations, or that of filters. During World War II, he participated in the Manhattan project, which led to the building of the atomic bomb.

[8] Robert G. GALLAGER (1931–) is an American scientist who is a specialist in information theory and telecommunications. Professor at MIT, he has also been president of the IEEE Information Theory Society.

[9] Claude BERROU (1951–) is a French scientist who is a specialist in electrical engineering and a professor at IMT Atlantique. We owe him the invention of turbocodes.

Exercises

Exercise 2.1. [p. 171] We consider a binary code on a 2-element set for which the two code words are 0 and 01.

a) Is it a prefix code?

b) Prove that this code is uniquely decodable.

Exercise 2.2. [p. 171] Let (X_n) be a stationary random process with values in a finite set A. For any integer n, we consider the random variable $Y_n = (X_1, \ldots, X_n)$ with values in A^n.

a) Assuming moreover that the X_n are independent, what is the law of Y_n? What is its entropy?

b) Let C_n be a code on A^n with values in an alphabet of cardinality D, and which is optimal with respect to the law of Y_n. For $n > 0$, we set

$$L_n = \mathbf{E}\Big(\ell(C_n(Y))\Big)/n.$$

Applying Shannon's theorem to the variable Y_n, prove that the sequence (L_n) converges to the entropy rate of the process (X_n).

Exercise 2.3. [p. 172] We consider the alphabet A = {a, b, c, d, e, f, g} with respective probabilities

a	b	c	d	e	f	g
0.49	0.26	0.12	0.04	0.04	0.03	0.02

a) Compute the entropy of a random variable that follows this law.

b) Using Shannon's method, construct an associated binary code, where each symbol $x \in A$ is coded by a word of length $\lceil -\log(p(x)) \rceil$, where $p(x)$ is its probability. What is the average length of this code?

c) Using Huffman's method, construct an optimal binary code. What is its average length?

d) Code the word *cabbage*.

e) Decode the word 111110111101111011101.

Exercise 2.4. [p. 174] We consider a random variable X with values in a finite alphabet A, following a probability law p. To compute an optimal binary code with respect to this law, we need to know the law p. Assume that we only know some approximation q and that we use a binary code C constructed by Shannon's method with respect to the law q; in particular, we have $\ell(C(a)) = \lceil -\log_2(q(a)) \rceil$.

a) We assume that q is *dyadic*, that is, for each $a \in A$, the probability $q(a)$ is a real number of the form $1/2^n$, for some integer n. Prove that C is an optimal code with respect to the law q. What is the average length $\mathbf{E}(\ell(C(X)))$ of a code word?

b) In the general case, establish the following bounds for this average length:

$$H(X) + D(p,q) \leq \mathbf{E}(\ell(C(X))) < H(X) + D(p,q) + 1.$$

Exercise 2.5. [p. 175] We consider a random variable X with values in the alphabet A = {a, b, c, d}. The following table indicates the law of X, as well as two binary codes on this alphabet.

X	p_X	code I	code II
a	0.4	1	1
b	0.3	01	10
c	0.2	001	100
d	0.1	0001	1000

For each of these two codes, answer the following four questions.

a) Is this a prefix code? Is it uniquely decodable?

b) Consider the random variables

U = "X = a"   and   V = "the first symbol of the code word is 1"

with values in true/false. What is their mutual information I(U, V)?

2.6. Coding adapted to a transmission channel. 91

c) What is the average length of a code word? Compare with Shannon's theorem.

d) Construct an optimal binary code for the random variable U. What is the average length of a code word?

Exercise 2.6. [p. 176]   *a)* Let X be a discrete random variable whose set of possible values is bounded. Prove that $\mathbf{P}(X \geq a) \leq \mathbf{E}(e^{tX})e^{-ta}$, for all $a \in \mathbf{R}$ and all $t > 0$. (*Chernoff's inequality*)

Let $(X_k)_{1 \leq k \leq n}$ be an independent sequence of random variables which follow a Bernoulli law with some common parameter p.

b) We set $S = (X_1 + \cdots + X_n)/n$. Compute $\mathbf{E}(e^{tS})$ for all $t > 0$.

c) Compute the divergence $D(q, p)$ of a Bernoulli variable of parameter p with respect to a Bernoulli variable of parameter q.

d) Let $q \in [0; 1]$ be such that $q > p$. Prove that $\mathbf{P}(S \geq q) \leq e^{-nD(q,p)}$.

e) Let $q \in [0; 1]$ be such that $q < p$. Prove that $\mathbf{P}(S \leq q) \leq e^{-nD(q,p)}$.

Exercise 2.7. [p. 177]   Let $(X_1, \ldots, X_n)$ be an independent sequence of random variables, all of them following a Bernoulli law with parameter p. Let $S = X_1 + \cdots + X_n$.

a) Let $D(q, p)$ be the divergence of a Bernoulli variable with parameter p with respect to a Bernoulli variable of parameter q (see exercise 2.6). By studying $D(q, p)$ as a function of p, prove *Pinsker's inequality:*

$$D(q, p) \geq \frac{1}{2}(q - p)^2.$$

(*Prove that $\partial_p^2 D(q, p) \geq 1$.*)

b) Using Chernoff's inequality, deduce the upper bound

$$\mathbf{P}(|S - pn| \leq \varepsilon pn) \leq 2 \exp\left(-\frac{1}{2}\varepsilon^2 p^2 n\right).$$

Exercise 2.8. [p. 178]   Let d be an integer such that ≥ 3. We consider a memoryless transmission channel C both of whose alphabets are equal to the finite group $A = \mathbf{Z}/d\mathbf{Z}$, with the following transmission probabilities:

$$p(a + 1 \mid a) = p(a - 1 \mid a) = p \quad \text{and} \quad p(a \mid a) = 1 - 2p,$$

where p is a real number such that $0 \leq p \leq 1/2$.

a) What is the transmission capacity of this channel?

b) Determine the law of a random variable X with values in A such that $I(C) = I(X, Y)$, where Y is a random variable X with values in A such that $\mathbf{P}(Y = b \mid X = a) = p(b \mid a)$ for any $a, b \in A$.

Exercise 2.9. [p. 179]   Let p be a real number such that $0 \leq p \leq 1$. Compute the transmission capacity of a memoryless transmission channel given by the following matrix of transmission probabilities: $\begin{pmatrix} 1 & 0 \\ p & 1-p \end{pmatrix}$, as well as the laws on the source that allow this capacity to be reached. Study how this transmission capacity varies with p; in particular, interpret the special cases $p = 1/2$, $p = 1$, $p = 0$.

Exercise 2.10. [p. 179] Let C' and C'' be memoryless transmission channels. We assume that the input alphabet of C'' is equal to the output alphabet of C', and we consider the transmission channel obtained by joining the two channels C' and C''.

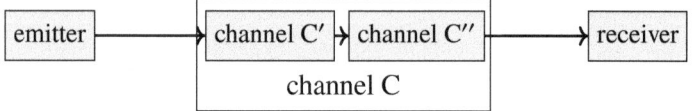

a) Compute the matrix of transmission probabilities of C in terms of those of C' and C''.

b) Prove that the transmission capacity of the channel C satisfies

$$I(C) \leq \inf\bigl(I(C'), I(C'')\bigr).$$

c) We join in this way n symmetric binary transmission channels with parameter p (that is, $p(1 \mid 0) = p(0 \mid 1) = p$). Prove that the resulting channel C_n behaves as a symmetric binary transmission channel with parameter $\bigl(1 - (1 - 2p)^n\bigr)/2$. What happens when $n \to +\infty$?

Exercise 2.11. [p. 181] Let C' and C'' be transmission channels, with input alphabets A' and A'', and output alphabets B' and B''. We consider the channel $C = C' \times C''$ with input alphabet $A = A' \times A''$, output alphabet $B = B' \times B''$ and transmission probabilities

$$p\bigl((b', b'') \mid (a', a'')\bigr) = p(b' \mid a')\, p(b'' \mid a'').$$

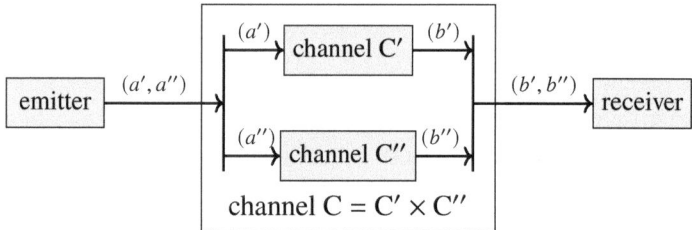

a) Let X', X'', Y', Y'' be random variable with values in A', A'', B', B'' respectively; we set $X = (X', X'')$ and $Y = (Y', Y'')$ and we assume that $X \sim_C Y$.
Prove that $X' \sim_{C'} Y'$ and $X'' \sim_{C''} Y''$.

b) Prove also that $X'' \perp_{X'} Y'$ and $Y'' \perp_{X''} X'$.

c) Prove that the transmission capacity of the channel C is given by

$$I(C) = I(C') + I(C'').$$

Determine a law on A that allows this capacity to be reached.

2.6. Coding adapted to a transmission channel.

Exercise 2.12. [p. 183]   Let C′ and C″ be transmission channels with input alphabets A′ and A″ and output alphabets B′ and B″. We assume that the sets A′ and A″ on the one hand, B′ and B″ on the other hand, are disjoint, and we set A = A′ ∪ A″ and B = B′ ∪ B″.

We consider the channel C on the input alphabet A with values in the alphabet B which has the transmission probabilities: $p_C(b \mid a) = p_{C'}(b \mid a)$ for $a \in A'$ and $b \in B'$, and $p_C(b \mid a) = p_{C''}(b \mid a)$ for $a \in A''$ and $b \in B''$.

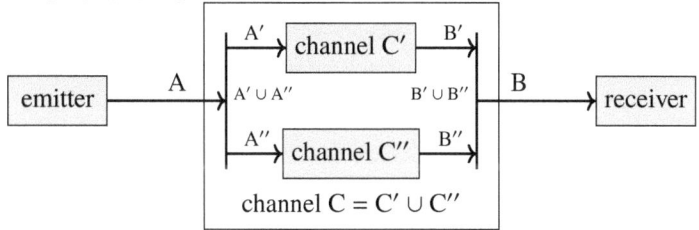

a) Explain why $p(b \mid a) = 0$ if either $b \in B'$ and $a \in A''$, or $b \in B''$ and $a \in A'$.

b) Prove the the transmission capacity of the channel C satisfies

$$e^{I(C)} = e^{I(C')} + e^{I(C'')}.$$

What probability laws on A allow this transmission capacity to be reached?

Exercise 2.13. [p. 185]   We consider a memoryless transmission channel C on alphabets A and B, but for which the use of a symbol $a \in A$ has a "cost" $c(a) \geq 0$. The capacity–cost function of this channel is defined by

$$I(C, \gamma) = \sup\{I(X, Y)\,;\, X \sim_C Y \text{ and } \mathbf{E}(c(X)) \leq \gamma\}.$$

a) Compute this function when C is a symmetric binary transmission channel with parameter p.

b) More generally, analyse this function when C is a channel on the alphabets A = B = $\{1, \ldots, d\}$ (with d symbols) whose matrix of transmission probabilities is given by

$$\begin{pmatrix} q & p & \cdots & p \\ p & q & \cdots & p \\ \vdots & \vdots & \ddots & \vdots \\ p & p & \cdots & q \end{pmatrix},$$

where p is a real number such $0 \leq p \leq 1/(d-1)$ and $q = 1 - (d-1)p$. (*It does not seem possible to give a closed form formula.*)

Exercise 2.14. [p. 188]   Let $p \in [0; 1/2[$. We consider the symmetric binary transmission channel C on the alphabets A = B = $\{0; 1\}$, with transmission probabilities $p(1 \mid 0) = p(0 \mid 1) = p$.

We also consider the code Φ of length 7 on the set M = A^4 with coding function f_Φ defined by

$$f_\Phi(m_1, m_2, m_3, m_4) = (m_1, m_2, m_3, m_4, c_1, c_2, c_3),$$

where

$$c_1 = m_2 + m_3 + m_4, \quad c_2 = m_1 + m_3 + m_4, \quad c_3 = m_1 + m_2 + m_3,$$

all additions being made modulo 2.

Once the source has given a block $m = (m_1, \ldots, m_4)$, the emitter gives the word $f_\Phi(m) = (m_1, \ldots, m_4, c_1, c_2, c_3)$ to the channel C, and the receiver gets a word $m' = (m'_1, \ldots, m'_7)$ and computes

$$e_1 = m'_2 + m'_3 + m'_4 - m'_5, \quad e_2 = m'_1 + m'_3 + m'_4 - m'_6, \quad e_3 = m'_1 + m'_2 + m'_3 - m'_7.$$

a) Assuming that no transmission error happened, check that $(e_1, e_2, e_3) = (0, 0, 0)$.

b) In this question, we assume that only one transmission error happened in the channel, at position i, that is, $m'_i \neq m_i$, but $m'_j = m_j$ for $j \neq i$. Compute (e_1, e_2, e_3) and explain how one can recover m.

Given the last two questions, we define a decoding function g_Φ as follows: we set $g_\Phi(m') = m$ when the words $(m, c) = f_\Phi(m)$ and m' differ by at most one symbol.

c) Assuming that m' differs from (m, c) by at least two symbols, prove that $m \neq g_\Phi(m')$.

d) What is the maximal probability of transmission error of this code?

Remark. — The code studied in this exercise was invented by Richard Hamming in 1950. Then an engineer at the Bell Telephone Laboratories, he had to program computers to make a computation during the weekend. He was however worried that an error in the data would prevent the machine from completing the computation correctly. Hamming told himself that if the machine could observe that an error had taken place, there should be a way to determine the location of this error, so that the computer could fix it by itself. His paper HAMMING (1950) set the ground for the theory of error correcting codes.

Chapter 3.
Sampling

This final chapter presents another aspect of information theory in Shannon's works, namely, the *sampling theorem* according to which one may only retain samples of a given signal, taken at regularly spaced instants, provided the sampling frequency is at least twice that of the frequencies that "appear" in the signal.

To give a rigorous mathematical meaning to this expression, we use the theory of Fourier[1] series and Fourier transformation. In a colorful way, we might say that this theory mirrors the time-view of a signal (what amplitude at what time?) and its frequency-view (what elementary sounds?). Combining these two points of view turned out to be extremely efficient and the theory of Fourier series is now important in all fields of mathematics, be they "pure" or "applied", as well as in many other fields of science.

Fourier series consider periodic signals that can be reconstructed by combining elementary signals with frequency a multiple of the fundamental frequency. The Fourier transformation handles more general signals; in this book, we shall make the hypothesis that their "energy" is finite, but the theory of tempered distributions would allow us to avoid this assumption.

While the framework of the first two chapters was probability theory, the present chapter belongs to analysis. To spare the readers any knowledge of Lebesgue theory (L^p-spaces, etc.), we made the choice to restrict most of the discussion to *piecewise continuous functions*, hence to neither give the most general statements, nor, at some points, complete proofs.

We can then prove the sampling theorem and also relate it to the Poisson summation formula.

---

[1] Joseph FOURIER (1768–1830) was a French mathematician, physicist and public servant. In his 1822 memoir devoted to the study of heat diffusion, he employed the method of decomposing a function as a sum of trigonometric functions, a method which now bears his name. We owe him many other contributions in mathematics (the beginning of linear programming) or physics (the discovery of the greenhouse effect). His scientific career followed the political instability of France at the beginning of the 19th century.

A last section studies the *uncertainty principle:* a signal cannot be simultaneously localized in time and in frequency. Well known in the context of quantum mechanics (Heisenberg), we shall see how this principle also touches information theory.

3.1. Continuous signals and discrete signals

Let us imagine a signal; it may be the singing of an artist, represented by the air pressure on the membrane of a microphone over time, or a visual signal, such as a view that we wish to photograph, represented by the luminosity emitted by each region of the scene. *Sampling* this signal means measuring it at various times or places, usually regularly spaced; this transforms our continuous signal (a function of time, or space), into a discrete signal (a sequence of values).

The sampling question was raised very early in communication theory. Already in the middle of the 19th century, engineers used it for example to transmit several signals on the same telegraph wire. Today, it is fundamental for the digital treatment of a signal, because computers can only handle a finite amount of information.

Actually, computer applications need to do more than sample a signal; they also have to *quantize* it, that is, transform continuous data (pressure, voltage...) into discrete data taking only finitely possible values (so as to be coded with 8 bits, 16 bits...). This combination of sampling and quantizing, and the later reconstruction of the signal, is at the heart of field signal processing in electrical engineering.

As its title indicates, we will only consider here the sampling question, by proving that we can reconstruct a sampled signal as soon as it does not contain any frequency greater than half of the sampling frequency. In other words, if we sample a signal to at least twice the frequencies it contains, we will be able to reconstruct it exactly, at least theoretically.

This mathematical theorem is commonly attributed to Shannon, because it seems to have been stated for the first time in SHANNON (1949a). However, Shannon insists there that it was "*common knowledge in the communication art.*" The idea of sampling at twice the frequency was also well known to H. Nyquist, and this theorem is also called the *Nyquist–Shannon theorem.*

Its proof relies on the theory of Fourier series and Fourier transformation that allows any signal to be decomposed into a combination of "pure" trigonometric signals. This theory will also bring a precise definition of the set of frequencies that appear in a given signal—it will be the support of its Fourier transform.

Considering such pure trigonometric signals already explains the necessity of sampling at the Nyquist[2] frequency. If we sample the function F defined by $F(t) = \sin(\omega t)$ (with period $2\pi/\omega$ and frequency $f = \omega/2\pi$) at the double frequency ω/π,

---

[2] Harry NYQUIST (1889–1976) was an American electronic and computer science engineer who worked all of his life at AT&T or the Bell Labs, both companies having in fact been founded by Alexander Graham Bell. A precursor to Shannon, Nyquist had shown in 1927 that it was necessary to sample at the double frequency. He also studied "thermal noise", which limits the sensibility of any electronic system such as a camera sensor.

3.1. Continuous signals and discrete signals.

we get the values $F(n\pi/\omega) = \sin(n\pi) = 0$, so that we can't distinguish the signal F from the zero signal!

We can also explain right away how this theorem, as applied to sound signals, intervenes in everyday life.

The simple sounds that we perceive, that of a singing voice, a musical instrument, etc., have a *fundamental frequency* f_1, that basically determines the note (A, B, C..., the A above middle C usually corresponding to the 440 Hz frequency) that we relate this sound to; when this fundamental frequency is doubled, we hear the "same" note, but at one octave higher. What makes these sounds rich is that they "contain" *harmonics*, that is, frequencies which are multiples $f_2 = 2f_1$, $f_3 = 3f_1$... of the fundamental frequency. Our perception of sound unites all of these signals into a unique sound. Up to the (musically important) subtleties of scales, when the fundamental frequency f_1 corresponds to the note C, the next harmonics correspond, for $f_2 = 2f_1$, to the C of the next octave, for $f_3 = 3f_1$ to the G of this octave, for $f_4 = 4f_1 = 2f_2$ to the C of the next further octave, then for $f_5 = 5f_1$ to the E of that octave, and for $f_6 = 6f_1 = 2f_3$ to the G of that octave, one octave higher that the third harmonic $f_3 = 3f_1$. We guess here the importance of the major triad (C-E-G, for example) in classical music harmony, but also how additions of a seventh (in this case, B or B^b) enriches it so much.

We have collected in table 3.1 the ranges of fundamental frequencies (rounded up to the closest Hz) of a few musical instruments. This table does not take into account the harmonics which, as we said, are responsible for the nature of the sound; it seems that the first 10 harmonics are important. It also does not take into account the *anharmonic* frequencies that also exist, but at a much lesser level. (For some instruments, such as the toms of a drum kit, the anharmonic frequencies are important, so that we cannot easily assign a note to the sound of drums; moreover, the way the instrument is played, the kind of sticks that the musician uses, has an important impact on the sound and on its pitch.)

piano	27.5 Hz	4 186 Hz
saxophone – soprano	233 Hz	1480 Hz
alto	139 Hz	831 Hz
tenor	104 Hz	659 Hz
baritone	65 Hz	440 Hz
drum kit – cymbals	200 Hz	10 000 Hz
snare drum	240 Hz	6 000 Hz
toms	120 Hz	5 000 Hz
bass drum	60 Hz	4 000 Hz
voice – soprano	261 Hz	1 047 Hz
tenor	123 Hz	440 Hz
bass	82 Hz	300 Hz
violin	196 Hz	2 794 Hz

Fig. 3.1 Ranges of fundamental frequencies for some musical instruments

Sound perception also depends on the way our ear functions, from the ear drum that transforms the variations of air pressure into mechanical vibrations that it transmits to the cochlea, a small spiral-shape bone filled with a liquid where thousands of stereocilia respond to the various sound frequencies and transform them into electric stimuli on neurons. The human ear is sensitive to frequency signals that range from 20 Hz to 20 000 Hz; according to the page WIKIPEDIA (2022), ideal conditions allow a wider frequency range to be perceived, from 12 Hz to 28 000 Hz, but the part where human hearing is the most efficient lies between 2 000 and 5 000 Hz.

Consequently, the applications to sound signals may take as a Nyquist frequency any frequency greater than 40 000 Hz. The one chosen by the AUDIO-CD standard format is 44.1 kHz; theoretically, it allows the whole humanly audible spectrum of a signal to be recreated. Of course, if we are satisfied with a lesser quality signal, we can sample at a lower frequency, for example 8 000 Hz for basic telephones or for the VoIP (*Voice over IP*) protocol.

To conclude this technological introduction, let us recall that the signals must not only be sampled, but also quantized. As an example, the AUDIO-CD format codes any sample using 16 bits (2 bytes), hence, in principle, for a stereo signal, 176 kB of information per second. In fact, this standard format prescribes that six samples be regrouped into a 192 bit frame (24 bytes), to which are added 8 bytes of an error correcting code, and one control byte (*subcode*); finally, 33 bytes are coded for every 6 samples, hence 242 kB per second. Given these figures, the duration of a music CD is typically limited to roughly 1 hour. Compression algorithms may allow longer signals to be stored: some CDs do not contain a signal as described above, but music files compressed using the MP3 coding format.

Most of this chapter is devoted to describing the theory of Fourier series and Fourier transformation. Even if these are two examples of a more general theory that encompasses them both at the same time (harmonic analysis on locally compact abelian groups), pedagogical tradition and practical use command us to present them in succession.

3.2. The Fourier series of a periodic function

3.2.1. — The theory of Fourier series concerns the frequency analysis of periodic functions; it relies on the observation that a trigonometric function $t \mapsto \exp(i\omega t)$ is periodic of period T if and only if ω is an integer multiple of $2\pi/T$, and on the idea that "any" periodic function of period T can be described as a *sum* of such trigonometric functions.

The only hypothesis that a function should satisfy in order that we may form its Fourier series is to be *locally integrable* in the sense of Lebesgue[3], which will allow its integral to be computed on any bounded interval. Another framework,

---

[3] Henri-Léon LEBESGUE (1875–1941) was a French mathematician. His name is bound to measure theory, the field he invented at the beginning of the 1900s, following earlier work of Émile Borel. This theory allowed a final form to be given to the fundamental theorem of calculus relating differentiation

3.2. The Fourier series of a periodic function.

more restrictive but enough for many applications, is that of *piecewise continuous functions,* already mentioned in the introduction to this chapter.

3.2.2. — Let k be an integer ≥ 0. We say that a function f defined on a compact interval $[a;b]$ of $\mathbf{R}$ is *piecewise $\mathscr{C}^k$* if there exists a finite increasing sequence $(a_0, a_1, \ldots, a_n)$ of real numbers such that $a_0 = a$, $a_n = b$, and such that for any $p \in \{1, \ldots, n\}$, the function f is of class $\mathscr{C}^k$ on the open interval $]a_{p-1}; a_p[$ and admits, as well as all of its derivatives of order $\leq k$, a (right) limit at a_{p-1} and a (left) limit at a_p.

If f is defined on an arbitrary interval of $\mathbf{R}$, we say that it is *piecewise $\mathscr{C}^k$* if its restriction to any compact interval contained in its interval of definition is piecewise $\mathscr{C}^k$.

If f is a piecewise $\mathscr{C}^k$ function, we shall commit the abuse of notation of writing $f^{(k)}$ for its kth derivative; this function is defined almost everywhere, more precisely, outside of a set having finite intersection with any compact interval of its interval of definition.

Lemma 3.2.3. — *a) Let $f: [a;b] \to \mathbf{C}$ be a continuous and piecewise $\mathscr{C}^1$ function. We have*

$$\int_a^b f'(t)\, dt = f(b) - f(a).$$

b) Let $u, v: [a;b] \to \mathbf{C}$ be continuous and piecewise $\mathscr{C}^1$ functions. We have the formula of integration by parts:

$$\int_a^b u'(t)v(t)\, dt = \left[u(t)v(t)\right]_a^b - \int_a^b u(t)v'(t)\, dt.$$

Proof. — a) Let $(a_0, \ldots, a_n)$ be a finite increasing sequence such that $a = a_0$, $b = a_n$, and such that f is $\mathscr{C}^1$ on each interval $]a_{k-1}, a_k[$. If x and y are real numbers such $a_{k-1} < x \leq y < a_k$, we have

$$\int_x^y f'(t)\, dt = f(y) - f(x),$$

by the classic fundamental theorem of calculus. When x tends to a_{k-1} (while being $> a_{k-1}$), $f(x)$ tends to $f(a_{k-1})$, because f is continuous; similarly, when y tends to a_k (while being $< a_k$), $f(y)$ tends to $f(a_k)$. Consequently,

$$\int_{a_{k-1}}^{a_k} f'(t)\, dt = f(a_k) - f(a_{k-1}).$$

Then we have

---

and integration, but its impact has been much wider, because it provided the foundations of the modern formalism of probability theory. Lebesgue also studied subtle properties of Fourier series.

$$\int_a^b f'(t)\,dt = \sum_{k=1}^n \int_{a_{k-1}}^{a_k} f'(t)\,dt = \sum_{k=1}^n (f(a_k) - f(a_{k-1}))$$
$$= f(b) - f(a).$$

b) The proof of this formula is analogous. In a simpler way, we may observe that the function f defined by $f(t) = u(t)v(t)$ is also continuous and piecewise $\mathscr{C}^1$, and that $f'(t) = u'(t)v(t) + u(t)v'(t)$ for all $t \in [a;b]$ for all but finitely many $t \in [a;b]$. Applying *a)* to this function f, we recover formula *b)*. □

Definition 3.2.4. — *Let $f : \mathbf{R} \to \mathbf{C}$ be a locally integrable function, periodic with period* $T > 0$. *Its* Fourier coefficients *are the complex numbers:*

$$a_n(f) = \frac{2}{T} \int_0^T f(t) \cos(2\pi nt/T)\,dt, \qquad (3.2.4.1)$$

$$b_n(f) = \frac{2}{T} \int_0^T f(t) \sin(2\pi nt/T)\,dt, \qquad (3.2.4.2)$$

$$c_n(f) = \frac{1}{T} \int_0^T f(t) e^{-2i\pi nt/T}\,dt, \qquad (3.2.4.3)$$

for $n \in \mathbf{Z}$.

Since the involved functions are periodic of period T and are integrated over a full period, we may in fact integrate over any interval of length T. In particular, it is often useful to consider the centered interval $[-T/2; T/2]$.

These coefficients are not independent. Since the sine function is odd, and the cosine function is even, we have $a_{-n}(f) = a_n(f)$ and $b_{-n}(f) = -b_n(f)$ for any integer n; in particular, $b_0(f) = 0$. Decomposing the complex exponential as a sum of sine and cosine,

$$e^{-2i\pi nt/T} = \cos(2\pi nt/T) - i\sin(2\pi nt/T),$$

we get the relation

$$c_n(f) = \frac{1}{2}(a_n(f) - ib_n(f)) \qquad (3.2.4.4)$$

that can be reversed as

$$a_n(f) = c_n(f) + c_{-n}(f) \qquad (3.2.4.5)$$
$$b_n(f) = i(c_n(f) - c_{-n}(f)). \qquad (3.2.4.6)$$

Consequently, it will be enough to work with the coefficients c_n, which we refer to as the "complex Fourier coefficients", as opposed to the coefficients a_n and b_n, which we refer to as the "real Fourier coefficients", or the "sine/cosine Fourier coefficients". In certain cases, it may be more convenient to use one of these sets of coefficients instead of the other.

3.2. The Fourier series of a periodic function.

These coefficients are linear in f:

$$c_n(\lambda f) = \lambda c_n(f), \qquad c_n(f+g) = c_n(f) + c_n(g),$$

and similarly for the coefficients a_n and b_n.

The cosine and sine functions take real values, and complex conjugation turns $e^{-2i\pi nt/T}$ into $e^{2i\pi nt/T}$. It follows that we also have the relations

$$a_n(\overline{f}) = \overline{a_n(f)}, \qquad b_n(\overline{f}) = \overline{b_n(f)}, \qquad c_n(\overline{f}) = \overline{c_{-n}(f)}.$$

Let us denote by $\check{f}$ the function $t \mapsto f(-t)$; it is periodic, of period T. Performing the change of variables $t' = -t$ in the integrals that define its Fourier coefficients, we get

$$a_n(\check{f}) = a_n(f), \qquad b_n(\check{f}) = -b_n(f), \qquad c_n(\check{f}) = c_{-n}(f).$$

In particular, if f is even ($\check{f} = f$), then $b_n(f) = 0$ for all n: its sine Fourier coefficients vanish, and it is enough to compute its cosine Fourier coefficients. Similarly, if f is odd ($\check{f} = -f$), then $a_n(f) = 0$ for all n, its cosine Fourier coefficients vanish, and it is enough to compute its sine Fourier coefficients.

3.2.5. — Let us still consider a function $f : \mathbf{R} \to \mathbf{C}$, locally integrable, and periodic of period $T > 0$. Its *Fourier series* is the (infinite in both directions) series

$$\sum_{n \in \mathbf{Z}} c_n(f) e^{2\pi i n t/T}. \tag{3.2.5.1}$$

Its "real Fourier series" is the series

$$\frac{1}{2}a_0(f) + \sum_{n=1}^{\infty} \left(a_n(f) \cos(2\pi nt/T) + b_n(f) \sin(2\pi nt/T) \right). \tag{3.2.5.2}$$

Let us observe that they depend on t: these are series of functions. However, at this point, we do not claim any convergence property for these series; in fact, convergence does not always hold, let alone to $f(t)$.

For any integer $n \geq 0$, we write

$$S_n(f)(t) = \sum_{p=-n}^{n} c_p(f) e^{2\pi i p t/T}. \tag{3.2.5.3}$$

If we express the c_p-coefficients in terms of a_p and b_p, we get the formula

$$S_n(f)(t) = c_0(f) + \sum_{p=1}^{n} \Big(c_p(f)\big(\cos(2\pi pt/T) + i\sin(2\pi pt/T)\big) \\ + c_{-p}(f)\big(\cos(2\pi pt/T) - i\sin(2\pi pt/T)\big) \Big)$$

$$= \frac{1}{2}a_0(f) + \sum_{p=1}^{n}\Big((c_p(f) + c_{-p}(f))\cos(2\pi pt/\mathrm{T})$$
$$+ i(c_p(f) - c_{-p}(f))\sin(2\pi pt/\mathrm{T})\Big),$$

hence the equality

$$S_n(f) = \frac{1}{2}a_0(f) + \sum_{p=1}^{n}(a_p(f)\cos(2\pi pt/\mathrm{T}) + b_p(f)\sin(2\pi pt/\mathrm{T})), \quad (3.2.5.4)$$

that justifies the definition (3.2.5.2) of the real Fourier series of the function f.

Example 3.2.6. — One says that f is a *trigonometric polynomial* if it is a (finite) linear combination of functions, all of the form $t \mapsto e^{2i\pi pt/\mathrm{T}}$, in other words, if there exists an integer $\mathrm{N} \geqslant 0$ and a family $(c_p)_{-\mathrm{N} \leqslant p \leqslant \mathrm{N}}$ of complex numbers such that

$$f(t) = \sum_{p=-\mathrm{N}}^{\mathrm{N}} c_p e^{2i\pi pt/\mathrm{T}}.$$

Decomposing the complex exponential in terms of sine and cosine, this also amounts to the existence of two families of complex numbers, $(a_p)_{0 \leqslant p \leqslant \mathrm{N}}$ and $(b_p)_{1 \leqslant p \leqslant \mathrm{N}}$, such that

$$f(t) = \frac{1}{2}a_0 + \sum_{p=1}^{\mathrm{N}} a_p \cos(2\pi pt/\mathrm{T}) + b_p \sin(2\pi pt/\mathrm{T}).$$

The complex numbers a_p, b_p and c_p are related by the formulas $a_0 = 2c_0$, $a_p = c_p + c_{-p}$ and $b_p = i(c_p - c_{-p})$, for $p \in \{1, \ldots, \mathrm{N}\}$, in one direction, and $c_0 = \frac{1}{2}a_0$, $c_p = \frac{1}{2}(a_p - ib_p)$ and $c_{-p} = \frac{1}{2}(a_p - ib_p)$, for $p \in \{1, \ldots, n\}$ in the other.

Observe that its complex Fourier coefficients are precisely given by $c_n(f) = c_n$ for $|n| \leqslant \mathrm{N}$, all other coefficients vanishing. Similarly, one has $a_n(f) = a_{-n}(f) = a_n$ for $0 \leqslant n \leqslant \mathrm{N}$, $b_n(f) = -b_{-n}(f) = b_n$ for $1 \leqslant n \leqslant \mathrm{N}$, and all other coefficients vanish. For the proof, it suffices, by linearity, to treat the case where $f(t) = e^{2i\pi pt/\mathrm{T}}$. In this case, one has

$$c_n(f) = \frac{1}{\mathrm{T}}\int_0^{\mathrm{T}} e^{2i\pi pt/\mathrm{T}} e^{-2i\pi nt/\mathrm{T}}\,dt = \frac{1}{\mathrm{T}}\int_0^{\mathrm{T}} e^{2i\pi(p-n)t/\mathrm{T}}\,dt.$$

When $n = p$, we get

$$c_p(f) = \frac{1}{\mathrm{T}}\int_0^{\mathrm{T}} 1\,dt = 1,$$

while if $n \neq p$, we have

$$c_n(f) = \frac{1}{\mathrm{T}}\left[\frac{\mathrm{T}}{2i\pi(p-n)}e^{2i\pi(p-n)t/\mathrm{T}}\right]_0^{\mathrm{T}} = 0.$$

The formulas for $a_n(f)$ and $b_n(f)$ are immediate consequences.

3.3. The main theorems of the theory of Fourier series.

We also observe that for any integer n such that $n \geq N$, we have $S_n(f)(t) = f(t)$.

Example 3.2.7. — Let us assume that f is a $\mathscr{C}^1$ function on **R**, periodic of period T. In this case, its derivative f' is a continuous function, and it is periodic of period T, so that it admits Fourier coefficients. Using integration by parts, we shall relate them to the Fourier coefficients of f. Indeed, for any integer n, we can write

$$\begin{aligned}
c_n(f') &= \frac{1}{T} \int_0^T f'(t) e^{-2\pi i n t/T} \, dt \\
&= \frac{1}{T} \left[f(t) e^{-2\pi i n t/T} \right]_0^T - \frac{1}{T} \int_0^T f(t)(-2\pi i n/T) e^{-2\pi i n t/T} \, dt \\
&= \frac{2\pi i n}{T} \times \frac{1}{T} \int_0^T f(t) e^{-2\pi i n t/T} \, dt \\
&= \frac{2\pi i n}{T} c_n(f).
\end{aligned} \qquad (3.2.7.1)$$

For the a_n and b_n Fourier coefficients, we get

$$a_n(f') = -\frac{2\pi n}{T} b_n(f) \quad \text{and} \quad b_n(f') = \frac{2\pi n}{T} a_n(f). \qquad (3.2.7.2)$$

This formulas still hold when f is continuous and piecewise $\mathscr{C}^1$: this is proved in exactly the same way using integration by parts for continuous and piecewise $\mathscr{C}^1$ functions (lemma 3.2.3).

3.3. The main theorems of the theory of Fourier series

Now that we have defined the Fourier series of a periodic function, we need to study the question of its convergence. Already raised in Fourier's 1822 memoir, it would only receive a satisfactory solution in 1829, thanks to Dirichlet's work, a solution which shall be enough for the scope of this book. Only in the 20th century would a complete solution be found, via Kolmogorov's counterexample (1928) and the "almost everywhere convergence" theorems of Carleson (1966) and Hunt (1968).

Proposition 3.3.1 (Bessel[4] relation). — *Let f be a piecewise continuous function, periodic of period $T > 0$. For any integer $n \geq 0$, we have*

$$\frac{1}{T} \int_0^T |f(t) - S_n(f)(t)|^2 \, dt = \frac{1}{T} \int_0^T |f(t)|^2 \, dt - \sum_{k=-n}^{n} |c_k(f)|^2.$$

---

[4] Friedrich Wilhelm BESSEL (1784–1846) was a German mathematician, physician and astronomer. In particular, we owe him the first accurate measurement of the distance between the Sun and the Earth. In mathematics, his name remained attached to special functions that intervene in the

Proof. — To establish this relation, we go back to the definition of $S_n(f)$ as given by formula (3.2.5.3). Then, for any $t \in \mathbf{R}$, we have

$$|f(t) - S_n(f)(t)|^2 = \left| f(t) - \sum_{k=-n}^{n} c_k(f) e^{2i\pi kt/T} \right|^2$$

$$= f(t)\overline{f(t)} - \sum_{k=-n}^{n} f(t)\overline{c_k(f)} e^{-2i\pi kt/T}$$

$$- \sum_{k=-n}^{n} \overline{f(t)} c_k(f) e^{2i\pi kt/T} + \sum_{k,\ell=-n}^{n} c_k(f) \overline{c_\ell(f)} e^{2i\pi(k-\ell)t/T}.$$

Integrating this relation on the interval $[0; T]$, we obtain a decomposition of $\int_0^T |f(t) - S_n(f)(t)|^2 \, dt$ as the sum of the following four terms:

- The first is $\int_0^T |f(t)|^2 \, dt$;
- The second is

$$-\sum_{k=-n}^{n} \overline{c_k(f)} \int_0^T f(t) e^{-2i\pi kt/T} \, dt = -\sum_{k=-n}^{n} \overline{c_k(f)} \times T c_k(f)$$

$$= -T \sum_{k=-n}^{n} |c_k(f)|^2 ;$$

- The third, being conjugate to the preceding one, is also equal to $-T \sum_{k=-n}^{n} |c_k(f)|^2$;
- Finally, the last term is equal to

$$\sum_{k,\ell=-n}^{n} c_k(f) \overline{c_\ell(f)} \int_0^T e^{2i\pi(k-\ell)t/T} \, dt = T \sum_{k=-n}^{n} |c_k(f)|^2$$

since $\int_0^T e^{2i\pi mt/T} \, dt = 0$ for $m \neq 0$.

Then,

$$\frac{1}{T} \int_0^T |f(t) - S_n(f)(t)|^2 \, dt = \frac{1}{T} |f(t)|^2 \, dt - \sum_{k=-n}^{n} |c_k(f)|^2,$$

as was to be shown. □

Remark 3.3.2. — The analytic character of the preceding proof hides its geometric nature, which is none other than Pythagoras's theorem in an infinite-dimensional vector space. Indeed, the expression $\langle f, g \rangle = \frac{1}{T} \int_0^T \overline{f(t)} g(t) \, dt$ is a kind of (hermitian) scalar product on the space $\mathscr{C}_T$ of piecewise continuous and T-periodic functions: it is linear in g, semilinear in f (multiplying f by a complex number λ multiplies the

---

solution of the Laplace equation in cylindrical coordinates, as well as in the Bessel inequality that he discovered in 1828 for the estimation of a periodic movement.

3.3. The main theorems of the theory of Fourier series.

result by $\bar{\lambda}$), it is positive ($\langle f, f \rangle \geq 0$), so that the expression $\|f\| = \sqrt{\langle f, f \rangle}$ defines a kind of norm on that space, up to the fact that the equality $\|f\| = 0$ only implies that f vanishes outside of a finite subset of the interval $[0;T]$.

This scalar product takes complex values (this is what we mean by the *hermitian* adjective), and the geometric interpretation only considers its real part: if $\|f\| = \|g\| = 1$, the Cauchy–Schwarz inequality states $|\langle f, g \rangle| \leq \|f\| \times \|g\| \leq 1$, and the real part of this complex number $\langle f, g \rangle$ is the *cosine* of the angle $\theta(f, g)$ that the vectors f and g make.

The functions e_n, for $n \in \mathbf{Z}$, defined by $e_n(t) = e^{2i\pi nt/T}$ are elements of $(\mathscr{C}_T, \|\cdot\|)$, they have norm 1, and are pairwise *orthogonal*. The Fourier coefficients of $f \in \mathscr{C}_T$ can be interpreted as scalar products: $c_n(f) = \langle e_n, f \rangle$. Consequently, for every $n \in \mathbf{N}$, the function $S_n(f)$ is the orthogonal projection of f on the subspace W_n generated by the functions e_k, for $k \in \mathbf{Z}$ such that $-n \leq k \leq n$, hence $\langle f - S_n(f), S_n(f) \rangle = 0$.

Pythagoras's theorem furnishes the relations

$$\|S_n(f)\|^2 = \sum_{k=-n}^{n} |c_n(f)|^2$$

and

$$\|f\|^2 = \|f - S_n(f)\|^2 + \|S_n(f)\|^2,$$

hence the Bessel relation.

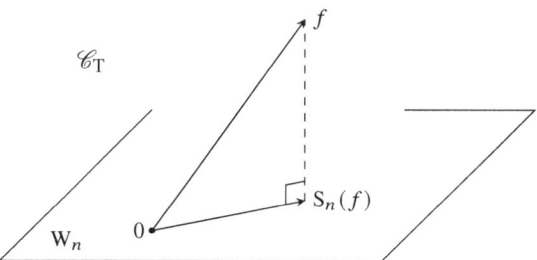

Fig. 3.2 Pythagoras's theorem in the space $\mathscr{C}_T$

We will make use of this geometric interpretation in the last section of this chapter (section 3.7).

Corollary 3.3.3 (Bessel inequality). — *Let f be a piecewise continuous function, periodic of period $T > 0$.*

$$\frac{1}{4}|a_0(f)|^2 + \frac{1}{2}\sum_{n=1}^{\infty}(|a_n(f)|^2 + |b_n(f)|^2) = \sum_{n \in \mathbf{Z}}|c_n(f)|^2 \leq \frac{1}{T}\int_0^T |f(t)|^2 \, dt.$$

We shall later prove Parseval's theorem (theorem 3.3.9), according to which the inequality is in fact an equality.

Proof. — We have $\frac{1}{4}|a_0(f)|^2 = |c_0(f)|^2$, as well as $\frac{1}{2}(|a_n(f)|^2 + |b_n(f)|^2) = |c_n(f)|^2 + |c_{-n}(f)|^2$. This implies the left equality. From the Bessel relation, we also deduce the inequality

$$\sum_{k=-n}^{n} |c_k(f)|^2 \leq \frac{1}{T}|f(t)|^2 \, dt,$$

valid for all integers $n \geq 0$. Letting n tend to $+\infty$, this implies the Bessel inequality.□

Since the general term of a converging series tends to 0, we have the following corollary:

Corollary 3.3.4. — *Let f be a piecewise continuous function, periodic of period $T > 0$. We have $\lim_{n \to \pm\infty} c_n(f) = 0$.*

Proposition 3.3.5. — *Let f be a continuous function, periodic with period $T > 0$. If all the Fourier coefficients of f vanish, then $f = 0$.*

Proof. — The function $t \mapsto f(2\pi t/T)$ is 2π-periodic and has the same Fourier coefficients (taking 2π as a period) as f. So we may assume that $T = 2\pi$.

Let us argue by contradiction. Let us assume that f is not identically zero, and let $u \in \mathbf{R}$ be such that $f(u) \neq 0$. The function $g: t \mapsto f(t-u)/f(u)$ is T-periodic, and its Fourier coefficients are given by

$$c_n(g) = f(u)^{-1} \int_0^T f(t-u) e^{-int} \, dt$$

$$= f(u)^{-1} e^{-inu} \int_{-u}^{T-u} f(t) e^{-int} \, dt$$

$$= f(u)^{-1} e^{-inu} c_n(f) = 0$$

for all integers n. Consequently, we can even assume that $u = 0$ and $f(0) = 1$. Since f is continuous, there exist real numbers $m > 0$ and $h \in [0; \pi/2]$ such that $f(t) \geq m$ for all $t \in [u-h; u+h]$.

For every integer $n \geq 0$, set $T_n(t) = (1 + \cos(t) - \cos(h))^n$; writing

$$1 + \cos(t) - \cos(h) = (1 - \cos(h)) + \frac{1}{2}e^{it} + \frac{1}{2}e^{-it},$$

we represent $T_n(t)$ in the form of a trigonometric polynomial: there are complex numbers m_k, with $-n \leq k \leq n$, such that $T_n(t) = \sum_{k=-n}^{n} m_k e^{ikt}$ for every $t \in \mathbf{R}$. Then

$$\int_{-\pi}^{\pi} T_n(t) f(t) \, dt = \sum_{k=-n}^{n} m_k \int_{-\pi}^{+\pi} e^{ikt} f(t) \, dt = \sum_{k=-n}^{n} m_k 2\pi c_{-k}(f) = 0.$$

For $t \in [-h; h]$, we have $\cos(t) \geq \cos(h)$, hence $1 + \cos(t) - \cos(h) \geq 1$, and then $T_n(t) \geq 1$. For $t \in [-h/2; h/2]$, we even have

3.3. The main theorems of the theory of Fourier series.

$$1 + \cos(t) - \cos(h) \geq 1 + \cos(h/2) - \cos(h) > 0.$$

On the other hand, for $t \in [-\pi; \pi] - [-h; h]$, we have $-1 \leq \cos(t) \leq \cos(h)$, hence $-1 \leq 1 + \cos(t) - \cos(h) \leq 1$ and $|T_n(t)| \leq 1$. This implies the following inequality

$$\left| \int_{-\pi}^{\pi} T_n(t) f(t) \, dt \right| \geq m \int_{-h}^{h} \left(1 + \cos(t) - \cos(h)\right)^n dt - \int_{-\pi}^{\pi} |f(t)| \, dt$$

$$\geq mh\left(1 + \cos(h/2) - \cos(h)\right)^n - \int_{-\pi}^{\pi} |f(t)| \, dt.$$

This expression tends to $+\infty$ when n tends to $+\infty$, while the left-hand side is identically 0. This is the desired contradiction. □

Corollary 3.3.6. — *Let f be a piecewise continuous function, periodic with period $T > 0$. If all of the Fourier coefficients of f vanish, then the set of all points $t \in [0; T]$ such that $f(t) \neq 0$ is finite.*

Using the periodicity of f, we can replace $[0; T]$ by any bounded interval.

A variant of this corollary holds when f is only assumed to be locally integrable; then the conclusion is that f is zero almost everywhere.

Proof. — Let $g \colon \mathbf{R} \to \mathbf{R}$ be the function $t \mapsto g(t) = \int_0^t f(s) \, ds$. For $t \in \mathbf{R}$, we have

$$g(t + T) - g(t) = \int_t^{t+T} f(s) \, ds = \int_0^T f(s) \, ds = T \times c_0(f) = 0.$$

Therefore, g is periodic with period T; it is continuous; it is also differentiable at any point t where f is continuous, with derivative $g'(t) = f(t)$, hence piecewise $\mathscr{C}^1$. By example 3.2.7, its Fourier coefficients are given by

$$c_n(g) = \frac{T}{2\pi i n} c_n(f) = 0,$$

for $n \in \mathbf{Z} - \{0\}$, so that the function $g^* \colon t \mapsto g(t) - c_0(g)$ is continuous and all its Fourier coefficients vanish. By the preceding theorem, we have $g^* \equiv 0$, so that g is constant, and identically 0 since $g(0) = 0$.

Since $g'(t) = f(t)$ at any point where f is continuous, this implies that $f(t) = 0$ for any point $t \in [0; T]$ except for finitely many $t \in [0; T]$. □

Corollary 3.3.7. — *Let f be a continuous function. If the Fourier series of f converges uniformly to some function g, then $f = g$.*

Proof. — A uniform limit of continuous functions is continuous, hence g is continuous. Let us prove that $c_n(g) = c_n(f)$ for every $n \in \mathbf{Z}$. Indeed, for any integer $m \geq |n|$, we have

$$c_n(f) = \frac{1}{T} \int_0^T S_m(f)(t) e^{-int} \, dt.$$

Since $\left(S_m(f)\right)_m$ converges uniformly to g, we then have

$$\frac{1}{T}\int_0^T S_m(f)(t)e^{-int}\,dt \to \frac{1}{T}\int_0^T g(t)e^{-int}\,dt = c_n(g).$$

This proves that $c_n(f) = c_n(g)$. Consequently, the continuous function $f-g$ satisfies $c_n(f-g) = c_n(f) - c_n(g) = 0$ for all $n \in \mathbf{Z}$. By the theorem, we then have $f - g \equiv 0$, hence $g = f$. □

Corollary 3.3.8. — *Let f be a continuous function, piecewise $\mathscr{C}^1$, periodic with period $T > 0$. The Fourier series of f converges uniformly to f.*

Proof. — If f is piecewise $\mathscr{C}^2$ and f, f' are continuous, we can use example 3.2.7 to estimate its Fourier coefficients: for every $n \in \mathbf{Z}$ such that $n \neq 0$, we have

$$c_n(f) = \frac{T}{2\pi i n} c_n(f') = \frac{T^2}{-4\pi^2 n^2} c_n(f'').$$

It follows from their definition as an integral that the Fourier coefficients of f'' are bounded by $\sup(|f''|)$. This implies the existence of a real number $A > 0$ such that $|c_n(f)| \leq A/n^2$, for all integers $n \in \mathbf{Z} - \{0\}$. From this upper bound and the convergence of the Riemann[5] series $\sum 1/n^2$, it follows that the Fourier series of f,

$$S_n(f)(t) = \sum_{k=-n}^{n} c_k(f) e^{ikt}$$

converges *normally*, in particular uniformly. By the preceding corollary, its limit is f.

When f is only assumed to be continuous and piecewise $\mathscr{C}^1$, we have to argue slightly differently. Let n and m be integers such that $n > m \geq 0$, and let $t \in \mathbf{R}$. We have

$$|S_n(f)(t) - S_m(f)(t)| = \left| \sum_{m+1 \leq |k| \leq n} c_k(f) e^{2i\pi kt/T} \right|$$

$$\leq \sum_{m+1 \leq |k| \leq n} |c_k(f)|$$

$$\leq \sum_{m+1 \leq |k| \leq n} |c_k(f')| \frac{T}{2\pi |k|}$$

$$\leq \frac{T}{2\pi} \left(\sum_{m+1 \leq |k| \leq n} |c_k(f')|^2 \right)^{1/2} \times \left(\sum_{m+1 \leq |k| \leq n} \frac{1}{k^2} \right)^{1/2}$$

$$\leq \frac{T}{2\pi} \left(\sum_{|k| \geq n} |c_k(f')|^2 \right)^{1/2} \times \left(2 \sum_{k=1}^{\infty} \frac{1}{k^2} \right)^{1/2},$$

---

[5] Bernhard RIEMANN (1826–1866) was a German mathematician. Besides groundbreaking contributions to geometry and complex analysis, we owe him the famous "Riemann hypothesis", a conjecture regarding the location of the complex zeros of the zeta function he formulated in his 1859 paper on the distribution of prime numbers. By the way, the sum of this series is $\pi^2/6$, see exercise 3.2.

3.3. The main theorems of the theory of Fourier series.

in which we used the Cauchy–Schwarz inequality. Since the series with general term $|c_k(f')|^2$ converges (Bessel's inequality, corollary 3.3.3), this upper bound implies that the sequence $(S_n(f))$ satisfies the Cauchy criterion for uniform convergence. Consequently, it converges uniformly; by the preceding corollary, its limit is the function f. □

Theorem 3.3.9 (Parseval[6] theorem). — *Let f be a piecewise continuous function, periodic with period $T > 0$. We have*

$$\frac{1}{T}\int_0^T |f(t)|^2\,dt = \sum_{n\in\mathbb{Z}} |c_n(f)|^2$$
$$= \frac{1}{4}|a_0(f)|^2 + \frac{1}{2}\sum_{n=1}^{\infty}(|a_n(f)|^2 + |b_n(f)|^2).$$

In fact, the formula also holds for more general functions than piecewise continuous functions, namely those which are locally square integrable in the sense of Lebesgue, for which the left-hand side can be defined, as well as the Fourier coefficients.

Given the Bessel relation, what Parseval's theorem proves is the following equality:

$$\lim_{n\to+\infty} \frac{1}{T}\int_0^T |f(t) - S_n(f)(t)|^2\,dt = 0.$$

In other words, the Fourier series of f converges "in quadratic mean" to f.

The proof of Parseval's theorem relies on the following approximation result.

Lemma 3.3.10. — *Let f be a piecewise continuous function, periodic with period $T > 0$. For every real number $\varepsilon > 0$, there exists a function g, continuous and piecewise $\mathscr{C}^1$, periodic with period T, such that*

$$\frac{1}{T}\int_0^T |f(t) - g(t)|^2\,dt \leq \varepsilon.$$

Proof. — Let α be a strictly positive real number which will be chosen later on. Let $(a_0,\ldots,a_n)$ be a strictly increasing sequence of real numbers such that $a_0 = 0$, $a_n = T$, and such that on any interval $]a_{k-1};a_k[$, the function f is the restriction of a continuous function f_k on $[a_{k-1};a_k]$. Since this interval is compact, the function f_k is uniformly continuous, by Heine's theorem, and there exists a real number $h > 0$ such that $|f_k(t) - f_k(s)| \leq \alpha$ if $|t-s| \leq h$. We now subdivide the interval $[a_{k-1};a_k]$ into m_k intervals of length $\leq h$, say $(b_{k,0}, b_{k,1},\ldots,b_{k,m_k})$, and consider the unique continuous function g_k on $[a_{k-1};a_k]$ which is affine on each subinterval $[b_{k,p-1};b_{k,p}]$, coincides with f at each of the points $b_{k,1},\ldots,b_{k,p-1}$,

---

[6] Marc-Antoine PARSEVAL (1755–1836) was a French mathematician who studied the partial differential equations governing the propagation of sound, and the evolution of fluids. In his 1799 paper, he stated a relation which is essentially equivalent to this formula. However, he does not give any proof, which he apparently considers as obvious, nor does he take convergence questions into account—they had to await Cauchy's work, around 1820.

equals $\frac{1}{2}\bigl(f_k(a_k) + f_{k+1}(a_k)\bigr)$ at a_k (for $k < n$; we set $g_n(a_n) = f_n(a_n)$ if $k = n$) and equals $\frac{1}{2}\bigl(f_k(a_{k-1}) + f_{k-1}(a_{k-1})\bigr)$ at a_{k-1} (for $k > 1$; we set $g_1(a_0) = f_1(a_0)$ of $k = 1$). Let M be an upper bound of $|f|$.

Let $t \in [0;\mathrm{T}]$. If t belongs to an interval of the form $[b_{k,p-1}; b_{k,p}]$, with $0 < p-1 < p < m_k$, then we have $|g_k(t) - f(t)| \leqslant \alpha$. If t belongs to an interval of the form $[b_{k,0}; b_{k,1}]$, or to an interval of the form $[b_{k,m_k-1}; b_{k,m_k}]$, then we can only assert that $|f_k(t) - f(t)| \leqslant 2\mathrm{M}$. Let g be the unique function on $\mathbf{R}$ which is periodic of period T and which coincides with g_k on $[a_{k-1}; a_k]$. We have

$$\frac{1}{\mathrm{T}} \int_0^\mathrm{T} |g(t) - f(t)|^2 \, dt \leqslant \alpha^2 + \frac{1}{\mathrm{T}} 8h\mathrm{M}^2.$$

Choosing $\alpha = \varepsilon/2$ and h small enough so that $h \leqslant \mathrm{T}\varepsilon/16\mathrm{M}^2$, we get the desired inequality. □

Proof (of Parseval's theorem). — If f is a periodic function, of period T, assumed to be, say, piecewise continuous, let us set

$$\|f\|_2 = \Bigl(\frac{1}{\mathrm{T}} \int_0^\mathrm{T} |f(t)|^2 \, dt\Bigr).$$

This is a seminorm on the space of such functions: it satisfies $\|af\|_2 = |a| \times \|f\|_2$ for all $a \in \mathbf{C}$ and any function f (homogeneity), and $\|f + g\| \leqslant \|f\| + \|g\|$ (triangle inequality). This is not a norm, because piecewise continuous function which vanish except on a finite number of points (per period) satisfy $\|f\|_2 = 0$, but these are the only ones.

For such a function f, let us also write

$$\|c(f)\|_2 = \Bigl(\sum_{k \in \mathbf{Z}} |c_k(f)|^2\Bigr)^{1/2}.$$

This is also a seminorm (homogeneity and triangle inequality); if $\|c(f)\|_2 = 0$, then $c_k(f) = 0$ for all $k \in \mathbf{Z}$ and this implies that f vanishes everywhere up to finitely many exceptions (per period).

Bessel's inequality states that $\|c(f)\|_2 \leqslant \|f\|_2$ for every such function f, and Parseval's theorem is the more precise equality: $\|c(f)\|_2 = \|f\|_2$. We shall prove it by approximating f by a function g which is continuous and piecewise $\mathscr{C}^1$. Let ε be a strictly positive real number; by lemma 3.3.10, there exists a function g, periodic of period T, continuous and piecewise $\mathscr{C}^1$, such that $\|f - g\|_2 \leqslant \varepsilon$; the triangle inequality then implies

$$\|g\|_2 \geqslant \|f\|_2 - \varepsilon.$$

By Bessel's inequality, we have $\|c(f - g)\|_2 \leqslant \|f - g\|_2 \leqslant \varepsilon$. Since $c_k(f) = c_k(f - g) + c_k(g)$ for all integers k, we deduce from the triangle inequality that

$$\|c(f)\|_2 \geqslant \|c(g)\|_2 - \|c(f - g)\|_2 \geqslant \|c(g)\|_2 - \varepsilon.$$

Since g is continuous and piecewise $\mathscr{C}^1$, the sequence of functions $(S_n(g))$ converges uniformly to g (corollary 3.3.8). Consequently, there exists an integer N such that $|g(t) - S_n(g)(t)| \leq \varepsilon$ for all $t \in \mathbf{R}$ and any integer $n \geq$ N. For such integers n, we have

$$\|g - S_n(g)\|_2 = \left(\frac{1}{T}\int_0^T |g(t) - S_n(g)(t)|^2 \, dt\right)^{1/2} \leq \varepsilon,$$

and Bessel's inequality implies that

$$\sum_{|k| \leq n} |c_k(g)|^2 \geq \frac{1}{T}\int_0^T |g(t)|^2 \, dt - \varepsilon^2.$$

When n tends to infinity, we have

$$\|c(g)\|_2^2 \geq \|g\|_2^2 - \varepsilon^2.$$

Finally,

$$\|c(f)\|_2 \geq \|c(g)\|_2 - \varepsilon \geq \left(\|g\|_2^2 - \varepsilon^2\right)^{1/2} \geq \left((\|f\|_2 - \varepsilon)^2 - \varepsilon^2\right)^{1/2}.$$

When ε tends to 0, we obtain $\|c(f)\|_2 \geq \|f\|_2$, which concludes the proof of Parseval's theorem. □

3.4. Convolution and Dirichlet's theorem

In this section, we present another approach to the theory of Fourier series and their convergence.

3.4.1. — Let f be a piecewise continuous function on $\mathbf{R}$. Let f^* be the function defined on $\mathbf{R}$ by the relation

$$f^*(t) = \frac{1}{2}\left(f(t^-) + f(t^+)\right),$$

where $f(t^+)$ and $f(t^-)$ are the right and left limit of f at t; we say that it is the *Dirichlet*[7] *regularization* of f.

If I is an open interval such that $f|_{\mathrm{I}}$ is continuous, then $f^* \equiv f$ on I. Consequently, the set of points where f and f^* differ meets any bounded interval in only finitely many points.

---

[7] Peter Gustav LEJEUNE DIRICHLET (1805–1859) was a German mathematician. In 1829, he was the first to establish in a rigorous way the convergence condition for Fourier series. In number theory, he proved the *arithmetic progression theorem*: every arithmetic progression whose first term and common difference are coprime contains infinitely many prime numbers. He also worked in mathematical physics.

Theorem 3.4.2. — *Let $f: \mathbf{R} \to \mathbf{C}$ be a piecewise $\mathscr{C}^1$-function, periodic with period T. For every $x \in \mathbf{R}$, we have*

$$S_n(f)(x) \to f^*(x)$$

when $n \to \infty$.

3.4.3. — The proof starts by rewriting $S_n(f)(x)$ as an integral:

$$S_n(f)(x) = \sum_{p=-n}^{n} c_p(f) e^{2\pi i p x/T}$$

$$= \sum_{p=-n}^{n} \left(\frac{1}{T} \int_0^T f(t) e^{-2\pi i p t/T} \, dt \right) e^{2\pi i p x/T}$$

$$= \frac{1}{T} \int_0^T f(t) \sum_{p=-n}^{n} e^{2\pi i p (x-t)/T} \, dt.$$

For any $u \in \mathbf{R}$, let us then set

$$D_n(u) = \frac{1}{T} \sum_{p=-n}^{n} e^{2\pi i p u/T} \, ; \qquad (3.4.3.1)$$

the function D_n is a trigonometric polynomial, we call it the *Dirichlet kernel*. By construction, we have

$$S_n(f)(x) = \int_0^T f(t) D_n(x-t) \, dt. \qquad (3.4.3.2)$$

By the change of variables $t' = x - t$ and the periodicity of f and D_n, we can rewrite this as

$$S_n(f)(x) = \int_{x-T}^{x} f(x-u) D_n(u) \, du = \int_{-T/2}^{T/2} f(x-u) D_n(u) \, du. \qquad (3.4.3.3)$$

Lemma 3.4.4. — *Let n be a natural number.*
1. *We have $\int_0^T D_n(u) \, du = 1$.*
2. *Let $u \in \mathbf{R}$. If $u \in T\mathbf{Z}$, then $D_n(u) = 2n + 1$; otherwise,*

$$D_n(u) = \frac{1}{T} \frac{\sin\left(\pi(2n+1)u/T\right)}{\sin(\pi u/T)}.$$

Proof. — The proof of the first relation is immediate. To prove the second one, we start from the definition of $D_n(u)$:

$$D_n(u) = \frac{1}{T} \sum_{p=-n}^{n} e^{2\pi i p u/T} = \frac{1}{T} e^{-2\pi i n u/T} \sum_{p=0}^{2n} e^{2\pi i p u/T}.$$

3.4. Convolution and Dirichlet's theorem.

We recognize the sum of the first $(2n + 1)$ terms of a geometric progression with common ratio $e^{2\pi i u/T}$. For $u \in T\mathbf{Z}$, all of these terms are equal to 1, hence $D_n(u) = 2n + 1$. Otherwise, we have $e^{2\pi i u/T} \neq 1$, so that

$$\begin{aligned}
D_n(u) &= \frac{1}{T} e^{-2\pi i n u/T} \frac{1 - e^{2\pi i (2n+1)u/T}}{1 - e^{2\pi i u/T}} \\
&= \frac{1}{T} e^{-2\pi i n u/T} \frac{e^{\pi i (2n+1)u/T}\bigl(-2i \sin(\pi(2n+1)u/T)\bigr)}{e^{\pi i u/T}\bigl(-2i \sin(\pi u/T)\bigr)} \\
&= \frac{1}{T} \exp\bigl((-2n + (2n+1) - 1)\pi i u/T\bigr) \frac{\sin\bigl(\pi(2n+1)u/T\bigr)}{\sin(\pi u/T)} \\
&= \frac{1}{T} \frac{\sin\bigl(\pi(2n+1)u/T\bigr)}{\sin(\pi u/T)}.
\end{aligned}$$

This proves the lemma. □

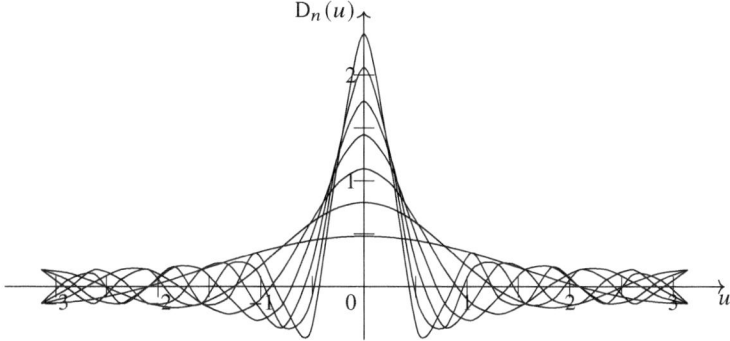

Fig. 3.3 Graphs of the Dirichlet kernel ($T = 2\pi$), for $1 \leq n \leq 7$

3.4.5. — The graph of the Dirichlet kernel on the interval $[-T/2; T/2]$ suggests that when n grows, this kernel is concentrated around 0 (modulo T). Together with the first relation of the preceding lemma, this could have been the key to the proof of Dirichlet's theorem, but as a consequence of the sign changes of D_n, one has

$$\lim_{n \to +\infty} \frac{1}{T} \int_0^T |D_n(u)|\, du = +\infty,$$

while $\frac{1}{T} \int_0^T D_n(u)\, du$ is constant, equal to 1.

So let us return to relation (3.4.3.3). Since D_n is even, we have

$$\int_0^{T/2} D_n(u)\, du = \int_{-T/2}^0 D_n(u)\, du = \frac{1}{2} \int_{-T/2}^{T/2} D_n(u)\, du = \frac{1}{2},$$

so that

$$S_n(f)(x) - f^*(x) = \int_{-T/2}^{T/2} f(x-t) D_n(t)\, dt$$
$$- \int_{-T/2}^{0} f(x^-) D_n(t)\, dt - \int_{0}^{T/2} f(x^+) D_n(t)\, dt$$
$$= \int_{0}^{T/2} \left(f(x-t) - f(x^-)\right) D_n(t)\, dt$$
$$+ \int_{-T/2}^{0} \left(f(x-t) - f(x^+)\right) D_n(t)\, dt$$
$$= \int_{0}^{T/2} \left(\left(f(x-t) - f(x^-)\right) + \left(f(x+t) - f(x^+)\right)\right) D_n(t)\, dt.$$

Since f is piecewise $\mathscr{C}^1$, the expressions $\frac{f(x-t)-f(x^-)}{t}$ and $\frac{f(x+t)-f(x^+)}{t}$ have a limit when t goes to 0. Consequently, there exists a continuous function g_x on $[0; T/2]$ such that
$$\left(f(x-t) - f(x^-)\right) + \left(f(x+t) - f(x^+)\right) = t g_x(t)$$
for every $t \in [0; T/2]$. We then write

$$S_n(f)(x) - f^*(x) = \int_{0}^{T/2} g_x(t) \frac{t}{T \sin(\pi t/T)} \sin\left(\pi(2n+1)t/T\right) dt.$$

The function defined by $t \mapsto t/\sin(\pi t/T)$ for $t \in {]}0; T/2]$ is continuous and has a limit at 0, namely T/π; consequently, it extends to a continuous function on $[0; T/2]$. Then the function on $]0; T/2]$ defined by
$$t \mapsto g_x(t) \frac{t}{T \sin(\pi t/T)}$$
for $t \in {]}0; T/2]$ extends to a continuous function h_x on $[0; T/2]$, and we have

$$S_n(f)(x) - f^*(x) = \int_{0}^{T/2} h_x(t) \sin\left(\pi(2n+1)t/T\right) dt. \tag{3.4.5.1}$$

By the following lemma (lemma 3.4.6), the right-hand side of this equation converges to 0 when $n \to +\infty$, so that
$$\lim_{x \to \infty} S_n(f)(x) = f^*(x).$$

This completes the proof of Dirichlet's theorem. □

3.4. Convolution and Dirichlet's theorem.

Lemma 3.4.6 (Riemann–Lebesgue lemma). — *Let $h\colon \mathbf{R} \to \mathbf{C}$ be a locally integrable function on $\mathbf{R}$ such that*

$$\int_{-\infty}^{\infty} |h(t)|\, dt < +\infty.$$

We have

$$\lim_{\omega \to \pm\infty} \int_{-\infty}^{+\infty} h(t) e^{-i\omega t}\, dt = 0.$$

In particular, this lemma holds for any piecewise continuous function which vanishes outside of a bounded interval of $\mathbf{R}$. In fact, we shall only give a complete proof in this particular case.

Proof. — To simplify the notation, let us set

$$\widehat{h}(\omega) = \int_{-\infty}^{\infty} h(t) e^{-i\omega t}\, dt,$$

for any function h as in the statement of the theorem, and $\omega \in \mathbf{R}$. Let us first assume that h is $\mathscr{C}^1$ on some compact interval $[a;b]$, and identically zero outside it. Then we can integrate by parts in the definition of $\widehat{h}(\omega)$, hence:

$$\widehat{h}(\omega) = \left[h(t)\frac{i}{\omega} e^{-i\omega t}\right]_a^b - \frac{i}{\omega} \int_a^b h'(t) e^{-i\omega t}\, dt.$$

This expression shows that we have

$$|\omega \widehat{h}(\omega)| \leqslant |h(a)| + |h(b)| + \int_a^b |h'(t)|\, dt,$$

so that $|\widehat{h}(\omega)| = O(1/\omega)$. In particular, $\lim_{\omega \to \pm\infty} \widehat{h}(\omega) = 0$.

In the general case, this argument is not enough since integration by parts does not hold. However, one can approximate h "in mean" by a step function which is zero outside of a bounded interval. Let us explain how this works under the assumption that h is piecewise continuous. Let $\varepsilon > 0$. Let us choose real numbers a and b such that $a < b$ and such that the two integrals

$$\int_{-\infty}^{a} |h(t)|\, dt \qquad \text{and} \qquad \int_{b}^{+\infty} |h(t)|\, dt$$

are smaller than ε. Let us then choose a finite increasing sequence $(a_0, \ldots, a_n)$ such that $a = a_0$ and $a_n = b$, and such that for any $k \in \{1, \ldots, n\}$, the function h is continuous on $]a_{k-1}, a_k[$ and has a right limit at a_{k-1}, and a left limit at a_k. Let h_k be the continuous function on $[a_{k-1}, a_k]$ that coincides with h on the open interval $]a_{k-1}, a_k[$. By Heine's theorem, these functions are uniformly continuous: there exists a real number $\delta > 0$ such that for all k and all $x, y \in [a_{k-1}, a_k]$ such that $|x - y| \leqslant \delta$, we have $|h_k(x) - h_k(y)| < \varepsilon/(b - a)$. Let us then subdivide the interval

$[a_{k-1}, a_k]$ into m_k subintervals of length $< \delta$, say $[a_{k,j-1}, a_{k,j}]$, for $1 \leqslant j \leqslant m_k$, with $a_{k,0} = a_{k-1}$ and $a_{k,m_j} = a_k$. Let g_k be the constant function on the interior of each of those intervals which coincides with h_k at their middle point $b_{k,j} = (a_{k,j-1} + a_{k,j})/2$. On any of those intervals I, we have $|g_k(x) - h_k(x)| < \varepsilon/(b-a)$, so that the integral of $|g_k - h_k|$ on I is smaller than $\varepsilon \ell(\mathrm{I})/(b-a)$. This implies the inequality $\int_{a_{k-1}}^{a_k} |g_k - h_k| < \varepsilon(a_k - a_{k-1})/(b-a)$.

Let g be a function on $\mathbf{R}$ which coincides with g_k on the interval $]a_{k-1}, a_k[$, for each k, and vanishes outside of $[a; b]$. This is a step function and we have

$$\int_{-\infty}^{\infty} |g - h| = \int_{-\infty}^{a} + \int_{a}^{b} + \int_{b}^{\infty} |g - h| \leqslant 3\varepsilon.$$

From this inequality, we deduce the upper bound

$$|\widehat{g}(\omega) - \widehat{h}(\omega)| = \left| \int_{-\infty}^{\infty} (g(t) - h(t)) e^{-i\omega t} \, dt \right| \leqslant \int_{-\infty}^{\infty} |g(t) - h(t)| \, dt \leqslant 3\varepsilon,$$

for all $\omega \in \mathbf{R}$.

On the other hand, we have

$$\widehat{g}(\omega) = \sum_{k=1}^{n} \sum_{j=1}^{m_k} h(b_{k,j}) \int_{a_{k,j-1}}^{a_{k,j}} e^{-i\omega t} \, dt = \frac{1}{\omega} \sum_{k=1}^{n} \sum_{j=1}^{m_k} i h(b_{k,j}) (e^{-i\omega a_{k,j}} - e^{-i\omega a_{k,j-1}}),$$

so that
$$\lim_{\omega \to +\infty} \widehat{g}(\omega) = 0.$$

In particular, there exists a real number W such that $|\widehat{g}(\omega)| \leqslant \varepsilon$ as soon as $|\omega| \geqslant \mathrm{W}$. For such a real number ω, we then have $|\widehat{h}(\omega)| \leqslant 4\varepsilon$, and this concludes the proof of the Riemann–Lebesgue lemma. □

Remark 3.4.7. — In undergraduate calculus, one is taught to define the sum of a series by considering the limit of its partial sums, but there are many other methods. Two of them fit the theory of Fourier series particularly well.

For any real number r such that $0 \leqslant r < 1$, let us define

$$\mathrm{P}_r(f)(x) = \sum_{k=-\infty}^{\infty} c_k(f) r^{|k|} e^{2\pi i k x/\mathrm{T}}. \tag{3.4.7.1}$$

Since the Fourier coefficients of f are bounded, the additional factor $r^{|k|}$ implies an upper bound for the terms of this (infinite in both directions) series, so that it converges normally and its limit is a continuous function of x. We can express $\mathrm{P}_r(f)$ in a similar way to the relation (3.4.3.3):

3.4. Convolution and Dirichlet's theorem.

$$P_r(f)(x) = \sum_{k=-\infty}^{\infty} \frac{1}{T}\left(\int_0^T f(t)e^{-2\pi ikt/T}\,dt\right) r^{|k|} e^{2\pi ikx/T}$$

$$= \int_0^T f(t) \frac{1}{T}\left(\sum_{k=-\infty}^{\infty} r^{|k|} e^{2\pi ik(x-t)/T}\right) dt$$

$$= \int_0^T f(t) K_r(x-t)\,dt,$$

where the function K_r, the *Poisson kernel*, is defined by

$$K_r(t) = \frac{1}{T} \sum_{k=-\infty}^{\infty} r^{|k|} e^{2\pi ikt/T}.$$

We may give a closed form expression for K_r:

$$K_r(t) = 1 + \sum_{k=1}^{\infty} r^k e^{2\pi ikt/T} + \sum_{k=1}^{\infty} r^k e^{-2\pi ikt/T}$$

$$= 1 + \frac{re^{2\pi it/T}}{1 - re^{2\pi it/T}} + \frac{re^{-2\pi it/T}}{1 - re^{-2\pi it/T}}$$

$$= \frac{1 - r^2}{1 - 2r\cos(2\pi t/T) + r^2}.$$

We observe in this formula that the Poisson kernel is positive and converges uniformly to 0 on any interval $[\delta; T - \delta]$, when $r \to 1^-$; its definition also shows that it is even and that its integral is 1. A classical argument then shows that $K_r(t)(x)$ converges to the Dirichlet regularization $f^*(x)$ of f at any point x such that f has a left and a right limit at x.

Indeed, as in the proof of Dirichlet's theorem, we write

$$P_r(f)(x) - f^*(x) = \int_0^{T/2} \Big(\big(f(x-t) - f(x^-)\big) + \big(f(x+t) - f(x^+)\big)\Big) K_r(t)\,dt.$$

Let ε be a real number such that $\varepsilon > 0$. There exists a real number such that $\delta > 0$ $|f(x-t) - f(x^-)| \leqslant \varepsilon$ and $|f(x+t) - f(x^+)| \leqslant \varepsilon$ for any t satisfying $0 < t < \delta$. Since $K_r(t)$ is positive, the restriction to $[0; \delta]$ of the preceding integral is bounded above by

$$\int_0^\delta 2\varepsilon K_r(t)\,dt \leqslant 2\varepsilon \int_0^{T/2} K_r(t)\,dt = \varepsilon.$$

If M is an upper bound of $|f|$, the remaining part of the integral is less than

$$4M \int_\delta^{T/2} K_r(t)\,dt,$$

an expression which tends to 0 when $r \to 1^-$ since $K_r(t)$ converges uniformly to 0 on the interval $[\delta; T/2]$. Then, for r close enough to 1, this integral is bounded above by ε, and $|P_r(f)(x) - f^*(x)| \leq 2\varepsilon$.

Another method consists in considering the Cesàro means of the sequence of partial sums $S_n(f)(x)$, setting

$$T_n(f)(x) = \frac{1}{n+1} \sum_{k=0}^{n} S_n(f)(x). \tag{3.4.7.2}$$

Let us set

$$F_n(t) = \frac{1}{n+1} \sum_{k=0}^{n} D_k(t);$$

this function F_n is called the *Fejér[8] kernel*, and one deduces from equation (3.4.3.3) that

$$T_n(f)(x) = \int_0^T f(t) F_n(x-t) \, dt.$$

The Fejér kernel is even, periodic with period T, and its integral is 1. Moreover, we have

$$F_n(t) = \frac{1}{n+1} \sum_{k=0}^{n} D_k(t)$$

$$= \frac{1}{(n+1)T \sin(\pi t/T)} \sum_{k=0}^{n} \sin\left(\pi(2n+1)t/T\right)$$

$$= \frac{1}{(n+1)T \sin(\pi t/T)^2} \sum_{k=0}^{n} \sin\left(\pi(2n+1)t/T\right) \sin(\pi t/T)$$

$$= \frac{1}{(n+1)T \sin(\pi t/T)^2} \times \sum_{k=0}^{n} \frac{1}{2}\Big(\cos\left(\pi(2n)t/T\right) - \cos\left(\pi(2n+2)t/T\right)\Big)$$

$$= \frac{1}{2(n+1)T \sin(\pi t/T)^2} \Big(1 - \cos\left(2\pi(n+1)t/T\right)\Big),$$

so that

$$F_n(t) = \frac{1}{(n+1)T} \left(\frac{\sin\left(\pi(n+1)t/T\right)}{\sin(\pi t/T)} \right)^2. \tag{3.4.7.3}$$

In particular, we observe that the Fejér kernel is *positive*. A variant of the argument that we used for the Poisson kernel then implies that if f has left and right limits at x, then $T_n(f)(x)$ converges to its Dirichlet regularization $f^*(x)$ at x.

---

[8] Lipót FEJÉR (1880–1959) was a Hungarian mathematician of Jewish origin (he changed his birth name, Weisz, to Fejér around 1900). A specialist in analysis, his theorem on the Cesàro summation of Fourier series, as well as the analog for the Laplace expansion of spherical harmonics, had a great impact on the analysis of orthogonal families and their use in mathematics and mathematical physics. He was dismissed by the fascist regime allied to the Nazis, he barely escaped assassination, but did not emigrate.

3.5. Fourier transformation

We now pass to the Fourier transformation which allows a frequency-based study of nonperiodic functions defined on $\mathbf{R}$.

3.5.1. — Let $f\colon \mathbf{R} \to \mathbf{C}$ be a locally integrable function such that $\int_{-\infty}^{\infty} |f(t)|\, dt < +\infty$. Its *Fourier transform* $\widehat{f}$ is defined by the formula

$$\widehat{f}(y) = \int_{-\infty}^{\infty} f(x) e^{-2\pi i x y}\, dx. \tag{3.5.1.1}$$

Example 3.5.2. — Let W be a real number $\geqslant 0$, and let $f\colon \mathbf{R} \to \mathbf{C}$ be the function defined by $f(x) = 1$ for $x \in [-W; W]$, and $f(x) = 0$ otherwise. For any $y \neq 0$, we have

$$\begin{aligned}
\widehat{f}(y) &= \int_{-\infty}^{\infty} f(x) e^{-2\pi i x y}\, dx = \int_{-W}^{W} e^{-2\pi i x y}\, dx \\
&= \left[\frac{1}{-2\pi i y} e^{-2\pi i x y} \right]_{-W}^{W} = \frac{e^{-2\pi i W y} - e^{2\pi i W y}}{-2\pi i y} \\
&= \frac{\sin(2\pi y W)}{\pi y} = 2W\,\mathrm{sinc}(2\pi y W),
\end{aligned}$$

where the function sinc, sometimes called the *cardinal sine* function, is defined by

$$\mathrm{sinc}(y) = \sin(y)/y \tag{3.5.2.1}$$

for $y \neq 0$, and $\mathrm{sinc}(0) = 1$. For $y = 0$, we find $\widehat{f}(y) = 2W$, so that this formula still holds.

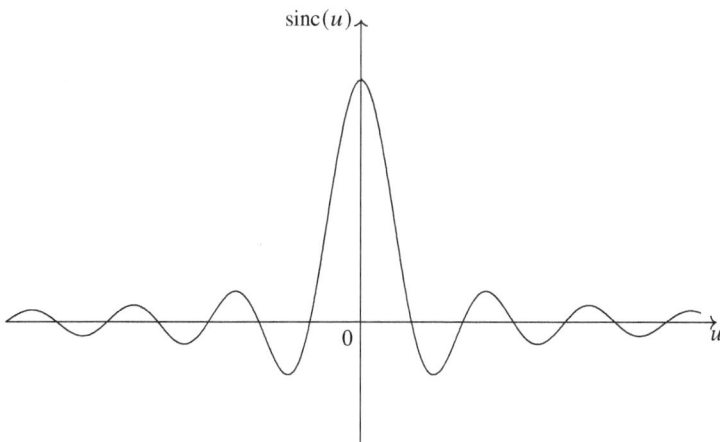

Fig. 3.4 Graph of the cardinal sine function

3.5.3. — The following properties immediately follow from the definition of the Fourier transform. Let f, g be locally integrable functions from $\mathbf{R}$ to $\mathbf{C}$ such that $\int_{-\infty}^{\infty} |f(x)|\, dx < +\infty$, and similarly for g.

1. We have $\widehat{f+g}(y) = \widehat{f}(y) + \widehat{g}(y)$ for all $y \in \mathbf{R}$.
2. For all $c \in \mathbf{C}$, we have $\widehat{cf}(y) = c\widehat{f}(y)$ for all $y \in \mathbf{R}$.
3. Let $a \in \mathbf{R}^*$ and $b \in \mathbf{R}$, and assume that $g(x) = f(ax+b)$ for all $x \in \mathbf{R}$; then we have $a\widehat{g}(ay) = e^{2\pi i b/a}\widehat{f}(y)$ for all $y \in \mathbf{R}$. Indeed,

$$\widehat{g}(y) = \int_{-\infty}^{\infty} f(ax+b)e^{-2\pi i x y}\, dx$$

$$= \int_{-\infty}^{\infty} f(ax+b) e^{-2\pi i (ax+b)(y/a)} e^{2\pi i b/a}\, dx$$

$$= \frac{1}{a} e^{2\pi i b/a} \int_{-\infty}^{\infty} f(x) e^{-2\pi i x(y/a)}\, dx$$

$$= \frac{1}{a} e^{2\pi i b/a} \widehat{f}(y/a).$$

4. Assume that $g(x) = \overline{f(x)}$; then $\widehat{g}(y) = \overline{\widehat{f}(-y)}$ for all $y \in \mathbf{R}$.

3.5.4. — Let f be a locally integrable function from $\mathbf{R}$ to $\mathbf{C}$ such that $\int_{-\infty}^{\infty} |f| < \infty$. We have $\widehat{f}(y) \leq \int_{-\infty}^{\infty} |f|$ for all $y \in \mathbf{R}$, and $\lim_{y \to \pm \infty} \widehat{f}(y) = 0$ (Riemann–Lebesgue lemma, lemma 3.4.6). Moreover, it follows from continuity theorems for parameterized integrals (relying on Lebesgue's dominated convergence theorem), that the function $\widehat{f}$ is continuous.

More generally, there is a kind of "correspondence" between on the one hand, decay properties for f and regularity properties for $\widehat{f}$, and on the other hand, regularity properties for f and decay properties for $\widehat{f}$.

1. Let p be a natural integer. If $x \mapsto x^p f(x)$ is integrable on $\mathbf{R}$, theorems for differentiation of parameterized integrals imply that $\widehat{f}$ is p-times continuously differentiable and that

$$\widehat{f}^{(p)}(y) = \int_{-\infty}^{\infty} (-2\pi i x)^p f(x) e^{-2\pi i x y}\, dx = \mathscr{F}\big((-2\pi i x)^p f\big)(y).$$

2. Let us moreover assume that f is continuous, piecewise $\mathscr{C}^1$, and that $\int_{-\infty}^{\infty} |f'| < \infty$. Then f has limits at $\pm\infty$, and these limits are necessarily 0. We can then integrate by parts in the definition of $\widehat{f'}$:

$$\widehat{f'}(y) = \int_{-\infty}^{\infty} f'(x) e^{-2\pi i x y}\, dx$$

$$= \left[f(x) e^{-2\pi i x y} \right]_{-\infty}^{\infty} + \int_{-\infty}^{\infty} f(x) 2\pi i y e^{-2\pi i x y}\, dx$$

$$= 2\pi i y \widehat{f}(y).$$

3.5. Fourier transformation.

Let us moreover assume that f is piecewise $\mathscr{C}^p$, that $f, \ldots, f^{(p-1)}$ are continuous, and that $\int_{-\infty}^{\infty} |f^{(k)}| < \infty$ for $k \in \{0, \ldots, p\}$. Then we have

$$\widehat{f^{(p)}}(y) = (2\pi i y)^p \widehat{f}(y),$$

for all $y \in \mathbf{R}$.

We denote by $\mathscr{S}(\mathbf{R})$ the set of all $\mathscr{C}^\infty$ functions from $\mathbf{R}$ to $\mathbf{C}$ such that for all integers p and k, the function $x \mapsto x^p f^{(k)}(x)$ is bounded on $\mathbf{R}$. This is a subspace of the space of functions from $\mathbf{R}$ to $\mathbf{C}$; we call it the *Schwartz class*. It is stable under multiplication by the function x, under differentiation, and, by the preceding computations, under Fourier transformation.

Theorem 3.5.5. — *Let $f: \mathbf{R} \to \mathbf{C}$ be a locally integrable function such that $\int_{-\infty}^{\infty} |f| < +\infty$.*

1. *If the function $\widehat{f}$ is integrable on $\mathbf{R}$, we have the* Fourier inversion formula

$$f(x) = \int_{-\infty}^{\infty} \widehat{f}(y) e^{2\pi i x y} \, dy,$$

for all $x \in \mathbf{R}$. In other words, $\widehat{\widehat{f}} = \check{f}$.

2. *The function $|f|^2$ is integrable on $\mathbf{R}$ if and only if $|\widehat{f}|^2$ is integrable on $\mathbf{R}$. In this case, we have* Plancherel's[9] *theorem:*

$$\int_{-\infty}^{\infty} |f(x)|^2 \, dx = \int_{-\infty}^{\infty} |\widehat{f}(y)|^2 \, dy.$$

Proof. — We shall only prove this theorem under the assumption that f is continuous, piecewise $\mathscr{C}^1$, and with compact support, that is to say, that there exists a real number T such that $f(x) = 0$ for $|x| \geq T/2$. Let then f_T be the function which coincides with f on $[-T/2; T/2]$ and which is periodic, of period T. It is continuous and piecewise $\mathscr{C}^1$. In particular, it is the sum of its Fourier series. Its Fourier coefficients can be computed as:

$$\begin{aligned} c_n(f_T) &= \frac{1}{T} \int_{-T/2}^{T/2} f(t) e^{-2\pi i n t/T} \, dt \\ &= \frac{1}{T} \int_{-\infty}^{\infty} f(t) e^{-2\pi i n t/T} \, dt \\ &= \frac{1}{T} \widehat{f}(n/T). \end{aligned}$$

For $x \in \mathbf{R}$ such that $|x| < T/2$, we then have

---

[9] Michel PLANCHEREL (1885–1967) was a Swiss mathematician, known to have extended the Fourier transformation to square integrable functions, a now fundamental tool in the field of partial differential equations.

$$f(x) = f_T(x) = \sum_{n \in \mathbf{Z}} c_n(f) e^{2\pi i n x/T} = \frac{1}{T} \sum_{n \in \mathbf{Z}} \widehat{f}(n/T) e^{2\pi i x n/T}.$$

In the right-hand side, we recognize a Riemann sum approximation of the integral $\int_{-\infty}^{\infty} \widehat{f}(y) e^{2\pi i x y} \, dy$, based on rectangles of width $1/T$. When $T \to +\infty$, it follows that

$$f(x) = \int_{-\infty}^{\infty} \widehat{f}(y) e^{2\pi i x y} \, dy = \widehat{\widehat{f}}(-x),$$

for all $x \in \mathbf{R}$. This establishes the Fourier inversion formula.

The proof of Plancherel's formula is analogous. We have

$$\int_{-\infty}^{\infty} |f(x)|^2 \, dx = \int_{-\infty}^{\infty} |f_T(x)|^2 \, dx$$
$$= T \sum_{n \in \mathbf{Z}} |c_n(f_T)|^2$$
$$= T \frac{1}{T^2} \sum_{n \in \mathbf{Z}} |\widehat{f}(n/T)|^2$$
$$= \frac{1}{T} \sum_{n \in \mathbf{Z}} |\widehat{f}(n/T)|^2,$$

the approximation of the integral $\int_{-\infty}^{\infty} |\widehat{f}(y)|^2 \, dy$ using Riemann sums based on rectangles of width $1/T$. When $T \to +\infty$, we get

$$\int_{-\infty}^{\infty} |f(x)|^2 \, dx = \int_{-\infty}^{\infty} |\widehat{f}(y)|^2 \, dy,$$

which proves Plancherel's theorem. □

Example 3.5.6. — Let us return to the function f from example 3.5.2, given by $f(x) = 1$ for $x \in [-W; W]$, and $f(x) = 0$ otherwise; we have $\int_{\mathbf{R}} |f(x)|^2 \, dx = 2W$. We have already seen that its Fourier transform is the function g given by $g(y) = 2W \sin(2\pi W y)/2\pi W y$ (and $g(0) = 1$). Consequently, $\int_{\mathbf{R}} |g(y)|^2 \, dy = 1$, so that

$$4W^2 \int_{\mathbf{R}} \left(\frac{\sin(2\pi W y)}{2\pi W y} \right)^2 \, dy = 2W.$$

Choosing $W = 1/2\pi$, we get

$$\int_{-\infty}^{\infty} \left(\frac{\sin(y)}{y} \right)^2 = 2\pi.$$

3.6. The sampling theorem

Theorem 3.6.1. — *Let $f: \mathbf{R} \to \mathbf{C}$ be a continuous function such that $x^2 f(x)$ is bounded, when x varies in $\mathbf{R}$. Then, f is integrable on $\mathbf{R}$.*

Assume moreover that there exists a real number W such that $\widehat{f}(y) = 0$ for all $y \in \mathbf{R}$ such that $|y| \geq W$. Then, we have, for any $x \in \mathbf{R}$, the relation

$$f(x) = \sum_{n=-\infty}^{\infty} f\left(\frac{n}{2W}\right) \operatorname{sinc}\left(x - \frac{n}{2W}\right).$$

In other words, if we sample a signal given by the function f at frequency 2W, we can reconstruct it exactly.

Proof. — Since f is continuous, integrable, and $\widehat{f}$ vanishes outside of the bounded interval $[-W; W]$, the Fourier inversion formula applies:

$$f(x) = \int_{-W}^{W} \widehat{f}(y) e^{2\pi i x y} \, dy.$$

By differentiation under the integration sign, this formula implies that f is indefinitely differentiable.

Let φ be the 2W-periodic function that coincides with $\widehat{f}$ on $[-W; W]$. Since $\widehat{f}(W) = \widehat{f}(-W) = 0$, it is continuous. Let us compute its Fourier coefficients.

For $n \in \mathbf{Z}$, we have

$$\begin{aligned} c_n(\varphi) &= \frac{1}{2W} \int_{-W}^{W} \varphi(y) e^{-2\pi i n y / 2W} \, dy \\ &= \frac{1}{2W} \int_{-\infty}^{\infty} \widehat{f}(y) e^{-2\pi i n y / 2W} \, dy \\ &= \frac{1}{2W} f(-n/2W). \end{aligned}$$

Consequently, the Fourier expansion of φ is given by

$$\varphi(y) = \sum_{n \in \mathbf{Z}} c_n(\varphi) e^{2\pi i n y / 2W} = \frac{1}{2W} \sum_{n \in \mathbf{Z}} f(n/2W) e^{-2\pi i n y / 2W}.$$

Since $n^2 f(n/2W)$ is bounded, for $n \in \mathbf{Z}$, this series converges uniformly when y varies. We can copy it in the Fourier inversion formula and integrate termwise: using that $\widehat{f}(y) = \varphi(y)$ for $y \in [-W; W]$ and $\widehat{f}(y) = 0$ otherwise, we get

$$f(x) = \frac{1}{2W} \sum_{n \in \mathbf{Z}} f(n/2W) \int_{-W}^{W} e^{2\pi i (x - n/2W) y} \, dy = \sum_{n \in \mathbf{Z}} f(n/2W) \operatorname{sinc}(x - n/2W),$$

as was to be shown. □

Remark 3.6.2. — According to SHANNON (1949a), the idea that one can reconstruct a signal by sampling it at a frequency at least twice the highest frequency that appears in this signal was well known to specialists in communication theory — he described it as "*common knowledge in the communication art*". The novelty of Shannon's paper, however, is to give an explicit formula for this reconstruction. SHANNON (1949a) admits that this formula had already been established by several mathematicians, in one form or another: let us quote, for example, E. Whittaker (1915), Oguro (1920), and Kotel'nikov (1933). Nevertheless, he insists that it appears for the first time in the context of communication theory.

This stammering piece of history may explain why the sampling theorem is sometimes referred to as the *Nyquist–Shannon theorem*.

Proposition 3.6.3. — *Let $f: \mathbf{R} \to \mathbf{C}$ be an integrable function; we assume that $\widehat{f}$ vanishes outside of an interval $[-W_0; W_0]$. Let W be a real number such that $W \geq W_0$, and let $\psi: \mathbf{R} \to \mathbf{C}$ be a piecewise continuous function which vanishes outside of $[-W; W]$ and which is identically equal to 1 on the interval $[-W_0; W_0]$. The, for every $x \in \mathbf{R}$, we have*

$$f(x) = \frac{1}{2W} \sum_{n \in \mathbf{Z}} f(n/2W) \widehat{\psi}(n/2W - x).$$

This formula is analogous to the reconstruction formula (3.6.1), which it recovers when $W = W_0$ and ψ is the indicator function of the interval $[-W_0; W_0]$. On the other hand, when $W > W_0$ (compared to the Nyquist $2W_0$, there is *oversampling*), we may take as the function ψ a $\mathscr{C}^\infty$ function. Then its Fourier transform $\widehat{\psi}$ decays rapidly at infinity, so that the given series converges very fast.

Proof. — Let us still denote by φ the 2W-periodic function which coincides with $\widehat{f}$ on $[-W; W]$; it is piecewise $\mathscr{C}^1$ and its Fourier coefficients are given by

$$c_n(\varphi) = f(-n/2W)/2W,$$

as in the proof of theorem 3.6.1. For any $y \in \mathbf{R}$, we have

$$\widehat{f}(y) = \varphi(y)\psi(y),$$

since either $y \in [-W; W]$, $\psi(y) = 1$, and $\widehat{f}(y) = \varphi(y)$, or $\widehat{f}(y) = \psi(y) = 0$. Writing φ as the sum of its Fourier series, we then have

$$\widehat{f}(y) = \sum_{n \in \mathbf{Z}} \frac{1}{2W} f(n/2W) e^{-2i\pi n y} \psi(y).$$

Let us apply the Fourier inversion formula: we get

$$f(x) = \sum_{n \in \mathbf{Z}} \frac{1}{2W} f(-n/2W) \int_{-\infty}^{\infty} e^{-2i\pi n y/2W} \psi(y) e^{2i\pi xy} \, dy,$$

3.6. The sampling theorem.

in which the integral is equal to $\widehat{\psi}(-x + n/2W)$. This concludes the proof. □

As a matter of fact, the proof of the sampling theorem is close to another fundamental result, the *Poisson*[10] *summation formula*.

Theorem 3.6.4 (Poisson summation formula). — *Let* $f : \mathbf{R} \to \mathbf{C}$ *be a function from the Schwartz class. Then, for any real number* $a > 0$, *we have*

$$\sum_{m \in \mathbf{Z}} f(am) = \frac{1}{a} \sum_{n \in \mathbf{Z}} \widehat{f}(n/a). \qquad (3.6.4.1)$$

More generally, for any real number $a > 0$ *and any real number* x, *we have*

$$\sum_{m \in \mathbf{Z}} f(x + am) = \frac{1}{a} \sum_{n \in \mathbf{Z}} \widehat{f}(n/a) e^{2i\pi nx/a}. \qquad (3.6.4.2)$$

The hypothesis that f belongs to the Schwartz class is stronger than necessary; it is enough that f be $\mathscr{C}^1$, and that f and f' decay sufficiently fast at infinity, for example, that there exist real numbers c, c' such that $|f(x)| \leqslant c/(1 + |x|^2)$ and $|f'(x)| \leqslant c'/(1 + |x|^2)$, for all x.

Proof. — Let us consider the series

$$F(x) = \sum_{m \in \mathbf{Z}} f(x + am).$$

The decay assumption on f and f' implies that this series converges, as well as its termwise derivative, uniformly when x belongs to a bounded interval. Consequently, this defines a $\mathscr{C}^1$ function on $\mathbf{R}$; by construction, this function is periodic with period a. Let us compute its Fourier coefficients: for every $n \in \mathbf{Z}$, we have

$$c_n(F) = \frac{1}{a} \int_0^a F(x) e^{-2i\pi nx/a}\,dx$$
$$= \frac{1}{a} \int_0^a \sum_{m \in \mathbf{Z}} f(x + ma) e^{-2i\pi nx/a}\,dx.$$

By uniform convergence of this series for $x \in [0; a]$, we may exchange integration and sum, so that

$$c_n(F) = \frac{1}{a} \sum_{m \in \mathbf{Z}} \int_0^a f(x + ma) e^{-2i\pi nx/a}\,dx.$$

We then get

---

[10] Siméon Denis PoissoN (1781–1840) was a French mathematician and physicist. One of the founders of the theory of electricity and magnetism, he brought substantial contributions to astronomy and mechanics. His works on integration and Fourier series were forerunners of those of Dirichlet and Riemann. He also wrote a book on probability and its application to court judgements.

$$c_n(F) = \frac{1}{a} \sum_{m \in \mathbf{Z}} \int_{ma}^{(m+1)a} f(x) e^{-2i\pi n(x-ma)/a} \, dx$$

$$= \frac{1}{a} \sum_{m \in \mathbf{Z}} \int_{ma}^{(m+1)a} f(x) e^{-2i\pi nx/a} \, dx$$

$$= \frac{1}{a} \int_{-\infty}^{\infty} f(x) e^{-2i\pi nx/a} \, dx = \widehat{f}(n/a).$$

Since F is $\mathscr{C}^1$, it is the sum of its Fourier series, hence the relation

$$\sum_{m \in \mathbf{Z}} f(x + am) = \frac{1}{a} \sum_{n \in \mathbf{Z}} \widehat{f}(n/a) e^{2i\pi nx/a},$$

for every $x \in \mathbf{R}$, which is the second formula of the theorem. By taking $x = 0$, we derive the first one. □

3.7. The uncertainty principle in communication theory

The following inequality is the archetypal uncertainty principle.

Proposition 3.7.1 (Heisenberg[11] inequality). — *Let $f : \mathbf{R} \to \mathbf{C}$ be a locally integrable function such that $|f|^2$ is integrable. We have*

$$\left(\int_{-\infty}^{\infty} x^2 |f(x)|^2 \, dx \right) \times \left(\int_{-\infty}^{\infty} y^2 |\widehat{f}(y)|^2 \, dy \right) \geq \frac{1}{16\pi^2} \left(\int_{-\infty}^{\infty} |f(x)|^2 \, dx \right)^2.$$

In quantum mechanics, $|f(x)|^2$ represents the probability density for a particle at position x, while $|\widehat{f}(y)|^2$ represents the probability density of this particle with momentum y. If its average position and average momentum are zero, the left-hand side is the product of the variance of its position and the variance of its momentum and the Heisenberg inequality bounds this product from below by a constant $(1/16\pi^2)$; the general case is similar. One often says that it is impossible to know with precision the position and the moment of a particle, and it is in this framework that HEISENBERG (1927) had stated this uncertainty inequality. The proof of the mathematical statement was given by KENNARD (1927), but the one we will follow is essentially that of WEYL (1950).

Proof. — To simplify, we shall only prove this under the assumption that f is $\mathscr{C}^1$ and compactly supported. Integrating by parts, we have

---

[11] Werner HEISENBERG (1901–1976) was a German physicist, inventor of quantum mechanics. After he participated in the Nazi effort towards nuclear energy, apparently aiming rather to an atomic pile than to a bomb, he played an important role in the reconstruction of the German scientific landscape after World War II.

3.7. The uncertainty principle in communication theory.

$$\int_{-\infty}^{\infty} |f(x)|^2 \, dx = -\int_{-\infty}^{\infty} x(|f^2|')(x) \, dx$$

$$= \int_{-\infty}^{\infty} x\bigl(f'(x)\overline{f(x)} + f(x)\overline{f'(x)}\bigr) \, dx$$

$$\leq 2 \int_{-\infty}^{\infty} |x| \times |f'(x)| \times |f(x)| \, dx.$$

Let us apply the Cauchy–Schwarz inequality: we obtain

$$\int_{-\infty}^{\infty} |f(x)|^2 \, dx \leq 2 \left(\int_{-\infty}^{\infty} |x|^2 \times |f(x)|^2 \, dx \right)^{1/2} \times \left(\int_{-\infty}^{\infty} |f'(x)|^2 \, dx \right)^{1/2}.$$

Then, using Plancherel's theorem, we have

$$\int_{-\infty}^{\infty} |f'(x)|^2 \, dx = \int_{-\infty}^{\infty} |\widehat{f'}(y)|^2 \, dy$$

$$= \int_{-\infty}^{\infty} |2\pi i y \widehat{f}(y)|^2 \, dy$$

$$= 4\pi^2 \int_{-\infty}^{\infty} |y|^2 \times |\widehat{f}(y)|^2 \, dy.$$

Combining these inequalities, we obtain

$$\left(\int_{-\infty}^{\infty} |f(x)|^2 \, dx \right)^2 \leq 4 \int_{-\infty}^{\infty} |x| \times |f'(x)| \times |f(x)| \, dx$$

$$\leq 4 \left(\int_{-\infty}^{\infty} |x|^2 \times |f(x)|^2 \, dx \right) \times \left(\int_{-\infty}^{\infty} |f'(x)|^2 \, dx \right)$$

$$\leq 4 \left(\int_{-\infty}^{\infty} |x|^2 \times |f(x)|^2 \, dx \right) \times \left(4\pi^2 \int_{-\infty}^{\infty} |y|^2 \times |\widehat{f}(y)|^2 \, dy \right)$$

$$\leq 16\pi^2 \left(\int_{-\infty}^{\infty} |x|^2 \times |f(x)|^2 \, dx \right) \times \left(\int_{-\infty}^{\infty} |y|^2 \times |\widehat{f}(y)|^2 \, dy \right),$$

as was to be shown. □

3.7.2. — In the rest of this section, we describe variants of the uncertainty principle in the framework of communication theory. One motivation could be the sampling theorem: it applies to functions f whose Fourier transform is compactly supported, in other words, signals whose spectrum is localized, and the uncertainty principle implies that such functions cannot be compactly supported: a (nonzero) signal cannot be simultaneously localized in time and in frequency. More generally, we have the following proposition.

Proposition 3.7.3. — *Let $f : \mathbf{R} \to \mathbf{C}$ be a nonzero integrable function whose Fourier transform vanishes outside of the interval $[-W; W]$, where W is a positive real number. Then there does not exist any interval $[u; v]$, with $u < v$, on which f is identically zero.*

Proof. — We argue by contraposition, by considering a function f which is integrable on **R**, identically zero on such an interval $[u; v]$ and whose Fourier transform is supported on the interval $[-W; W]$; we are going to prove that f is identically zero.

By the Fourier inversion formula, we have, for any $x \in [u; v]$, the equality

$$f(x) = \int_{-W}^{W} \widehat{f}(y) e^{2\pi i x y} \, dy = 0,$$

as well as for all derivatives of f:

$$\int_{-W}^{W} \widehat{f}(y)(2\pi i y)^p e^{2\pi i x y} \, dy = 0,$$

for all integers $p \geq 0$. Let us now apply the Fourier inversion formula for an arbitrary real number x:

$$f(x) = \int_{-W}^{W} \widehat{f}(y) e^{2\pi i x y} \, dy = \int_{-W}^{W} \widehat{f}(y) e^{2\pi i (x-u) y} e^{2\pi i u y} \, dy.$$

If we expand the exponential into its Taylor series,

$$e^{2\pi i (x-u) y} = \sum_{p=0}^{\infty} \frac{1}{p!} (2\pi i y)^p (x-u)^p,$$

we obtain

$$f(x) = \sum_{p=0}^{\infty} \frac{1}{p!} (x-u)^p \int_{-W}^{W} \widehat{f}(y)(2\pi i y)^p e^{2\pi i u y} \, dy = 0.$$

This proves that f vanishes identically. □

3.7.4. — Let us take an integrable function $f : \mathbf{R} \to \mathbf{C}$ so that we may consider its Fourier transform, and assume that it is normalized by

$$\int_{-\infty}^{\infty} |f(x)|^2 \, dx = 1.$$

By Plancherel's formula, we also have

$$\int_{-\infty}^{\infty} |\widehat{f}(y)|^2 \, dy = 1.$$

An important theme of communication theory consists in isolating the part of the signal represented by f which is located in a (time) interval $[-a; a]$, and to set

$$U_a(f) = \int_{-a}^{a} |f(x)|^2 \, dx.$$

3.7. The uncertainty principle in communication theory.

We have $U_a(f) \in [0;1]$ and, in some sense, $U_a(f)$ represents the energy of the signal which is localized in the time interval $[-a;a]$. Similarly, we can isolate the part of the signal whose spectrum is located in a frequency interval $[-b;b]$ and set

$$V_b(f) = \int_{-b}^{b} |\widehat{f}(y)|^2 \, dy.$$

Then $V_b(f) \in [0;1]$ and $V_b(f)$ represents the energy of the signal caused by the frequencies in the interval $[-b;b]$. By the preceding proposition, we cannot have $U_a(f) = V_b(f) = 1$.

Theorem 3.7.5 (Slepian, Landau, Pollak). — *There exists a real number $\theta > 0$ (depending on a and b) such that for any function $f \in L_2$, we have*

$$\arccos\left(\sqrt{U_a(f)}\right) + \arccos\left(\sqrt{V_b(f)}\right) \geq \theta.$$

The proof requires slightly more functional analysis that we intend to use in this book, so we shall only sketch the main arguments, referring the readers to the original papers SLEPIAN & POLLAK (1961); LANDAU & POLLAK (1961) or to the book DYM & MCKEAN (2016), where they will even find other applications to information theory.

In this theorem, we have denoted by L_2 the space of all functions from **R** to **C** which are square integrable (neglecting functions which vanish almost everywhere); it is endowed with the scalar product defined by $\langle f, g \rangle = \int_{\mathbf{R}} \overline{f}(x) g(x) \, dx$, and thanks to which L_2 is a Hilbert space. We write $\|f\|_2 = (\langle f, f \rangle)^{1/2}$ for the corresponding norm, for $f \in L_2$. Let us also recall the Cauchy–Schwarz inequality:

$$|\langle f, g \rangle| \leq \|f\|_2 \times \|g\|_2.$$

Let A be the linear mapping corresponding to time truncation: for any function f,

$$A(f)(x) = \begin{cases} f(x) & \text{if } |x| \leq a, \\ 0 & \text{otherwise.} \end{cases}$$

It is an orthogonal ($A^* = A$) projector ($A \circ A = A$) and its image is the closed subspace V of all functions in L_2 which vanish (almost everywhere) outside of $[-a;a]$.

Similarly, let B be the linear mapping corresponding to frequency truncation: for any function f,

$$\widehat{B(f)}(y) = \begin{cases} \widehat{f}(y) & \text{if } |y| \leq b, \\ 0 & \text{otherwise.} \end{cases}$$

Given the Fourier inversion formula, we thus have

$$B(f)(x) = \int_{-b}^{b} \widehat{f}(y) e^{2\pi i x y} \, dy.$$

The map B is also an orthogonal projector; its image is the closed subspace W of L_2 consisting of functions whose Fourier transform vanishes (almost everywhere) outside of $[-b\,;b]$. These are very particular functions: since the integration range is limited to the bounded interval $[-b;b]$, we may differentiate under the integral sign in the Fourier inversion formula, so that W consists of indefinitely differentiable functions. Moreover, for any $g \in W$, there are inequalities

$$|g^{(k)}(x)| = \left|\int_{-b}^{b}(2\pi i y)^k \widehat{g}(y)e^{2\pi i xy}\,dy\right| \leq (2\pi b)^k \sqrt{2b}\|g\|,$$

for all $k \geq 0$,

Proposition 3.7.6. — *We have* $\|B \circ A\| < 1$.

The norm $\|B \circ A\|$ of the linear mapping $B \circ A$ is the smallest real number γ such that $\|B \circ A(f)\| \leq \gamma\|f\|$ for all $f \in L_2$. Since A and B are orthogonal projectors, their norms are 1, hence their composition has norm $\gamma \leq 1$. (Explicitly, we have $\|B \circ A(f)\| \leq \|A(f)\| \leq \|f\|$.) The point of the proposition is to state the strict inequality $\gamma < 1$.

Geometrically, it means that the subspaces V and W in L_2 form an angle of measure at least $\arccos(\gamma) > 0$. Indeed, let us consider $f \in V$ and $g \in W$. The angle $\theta(f,g)$ that these functions make in L_2 appears in the Cauchy–Schwarz inequality:

$$\mathrm{Re}(\langle f,g\rangle) = \cos\big(\theta(f,g)\big)\|f\| \times \|g\|.$$

Since $f \in V$ and $g \in W$, we have $f = A(f)$ and $g = B(g)$, so that

$$\langle f,g\rangle = \langle A(f), B(g)\rangle = \langle B \circ A(f), g\rangle.$$

By the Cauchy–Schwarz inequality, we thus have

$$\mathrm{Re}(\langle f,g\rangle) \leq |\langle f,g\rangle| \leq |\langle B \circ A(f), g\rangle|$$
$$\leq \|B \circ A(f)\| \times \|g\| \leq \gamma\|f\| \times \|g\|,$$

hence the inequality $\cos\big(\theta(f,g)\big) \leq \gamma$.

As a first step of the proof, let us prove that the angle $\theta(f,g)$ is nonzero; equivalently, we have $V \cap W = 0$. Indeed, it follows from proposition 3.7.3 that the zero function is the only function which has compact support together with its Fourier transform.

From the point of view of signal theory, the inequality $\gamma < 1$ has a very concrete consequence: if we successively apply a time truncation, and then a frequency truncation, something that is automatically done in real life since we only consider time-limited signals and only perceive a limited range of their frequencies, then the energy of the resulting signal is only a fraction (at most γ) of the energy of the initial signal.

Proof (of proposition 3.7.6). — Since $A(f) = f$ for all $f \in V$ and $\|A\| \leq 1$, we have

$$\gamma = \|B \circ A\| = \sup_{f \in V-\{0\}} \frac{\|B(f)\|}{\|f\|} = \|B|_V\|.$$

3.7. The uncertainty principle in communication theory.

By symmetry, we also have

$$\gamma = \sup_{g \in W - \{0\}} \frac{\|A(g)\|}{\|g\|} = \|A|_W\|.$$

Let $g \in W$. By Fourier inversion, we have

$$g(x) = \int_{\mathbf{R}} \widehat{g}(y) e^{2\pi i x y} \, dy = \int_{-b}^{b} \widehat{g}(y) e^{2\pi i x y} \, dy,$$

so that

$$A(g)(x) = \begin{cases} \int_{-b}^{b} \widehat{g}(y) e^{2\pi i x y} \, dy & \text{if } |x| \leq a, \\ 0 & \text{otherwise.} \end{cases}$$

Let us choose a sequence (g_n) of functions from W such that $\|g_n\| = 1$ and $\|A(g_n)\| \to \gamma$. We saw that the functions (g_n) and all of their derivatives are bounded on each bounded interval. Let us now apply the Ascoli theorem from functional analysis: it implies that we can extract from the sequence (g_n) a subsequence which converges uniformly, on any bounded interval, as well as all of its derivatives. To simplify the notation, we replace the initial sequence by such a subsequence, and denote its limit by g.

Since $\|g_n\| \leq 1$, we have

$$\int_{-a}^{a} g(x)^2 \, dx = \lim_{n \to +\infty} \int_{-a}^{a} g_n(x)^2 \, dx \leq 1.$$

Consequently, $\|g\| \leq 1$; in particular, $g \in L_2$.

In fact, g belongs to the subspace W. Indeed, let us consider a function $h \in L_2$ which vanishes on $[-b; b]$. We then have $h\widehat{g_n} = 0$, for all n; by Plancherel's theorem, this implies

$$\langle \widehat{h}, g_n \rangle = \langle h, \widehat{g_n} \rangle = 0.$$

If $\widehat{h}$ is integrable on $\mathbf{R}$, then Lebesgue's dominated convergence theorem implies that the left-hand side converges to $\langle \widehat{h}, g \rangle$, so that $\langle \widehat{h}, g \rangle = 0$, hence $\langle h, \widehat{g} \rangle = 0$. Since h is arbitrary, we conclude that $\widehat{g}$ vanishes outside of $[-b; b]$, as was to be shown.

Since (g_n) converges uniformly to g on $[-a; a]$, we have $\|A(g)\| = \lim \|A(g_n)\| = \gamma$. Since $\|A(g)\| \leq \gamma \|g\|$, this implies $\|g\| = 1$.

Let us then prove that $B \circ A(g) = \gamma g$. Let $h \in W$. For $t \in \mathbf{C}$, the inequality $\|A(g + th)\|^2 \leq \gamma \|g + th\|^2$ writes as

$$\operatorname{Re}\left(t(\langle Ag, Ah \rangle - \gamma \langle g, h \rangle)\right) \leq t^2 (\|\gamma\| h\|^2 - \|A(h)\|^2).$$

For t close to 0, it implies the *equality* $\langle A(g), A(h) \rangle = \gamma \langle g, h \rangle$. Consequently, for all $h \in L_2$, we have

$$0 = \langle A(g) - \gamma g, B(h) \rangle = \langle B \circ A(g) - \gamma g, h \rangle,$$

hence $B \circ A(g) = \gamma g$.

Since $\|A(g)\| \leq 1$, the equality $\gamma = 1$ would imply that $\|A(g)\| = 1$, hence $A(g) = g$ and $g \in V$. However, we have proved that $V \cap W = 0$. Consequently, $\gamma < 1$. □

Proof (of theorem 3.7.5). — This statement may look complicated, but it results from a simple trigonometry argument in the subspace of L_2 spanned by the three vectors f, $A(f)$ and $B(f)$.

Let $f \in L_2$ be such that $\|f\| = 1$. Let $\alpha \in [0; \pi/2]$ be the angle formed by the vectors f and $A(f)$. Then $\|A(f)\| = \|f\| \cos(\alpha)$, hence $\|A(f)\|^2 = U_a(f) = \cos^2(\alpha)$ and then $\alpha = \arccos\left(\sqrt{U_a(f)}\right)$. Similarly, the angle between the vectors f and $B(f)$ is equal to $\beta = \arccos\left(\sqrt{V_b(f)}\right)$. Consequently, the angle formed by the vectors $A(f)$ and $B(f)$ is at most $\alpha + \beta$. On the other hand, this angle θ satisfies

$$\mathrm{Re}(\langle A(f), B(f)\rangle) = \cos(\theta)\|A(f)\| \times \|B(f)\| \leq \gamma \|A(f)\| \times \|B(f)\|,$$

so that $\alpha + \beta \geq \theta \geq \arccos(\gamma)$. This concludes the proof. □

Remark 3.7.7. — In fact, LANDAU & POLLAK (1961) prove that the inequality of the theorem (with $\theta = \arccos(\|B \circ A\|)$) is a necessary and sufficient condition for the existence of a function $f \in L_2$ such that $\|f\| = 1$ with given $U_a(f)$, $V_b(f)$ in $[0; 1]$, with two exceptions: if $U_a(f) = 0$, one cannot have $V_b(f) = 1$, and if $U_a(f) = 1$, one cannot have $V_b(f) = 0$.

Exercises

Exercise 3.1. [p. 189] Let f be a 2π-periodic function that is identically equal to -1 on $]-\pi; 0[$ and to 1 on $]0; \pi[$.

a) Compute its Fourier coefficients.

b) Apply Dirichlet's theorem at 0, then at $\pi/2$.

c) Apply Parseval's theorem.

Exercise 3.2. [p. 190] Let f be the 2π-periodic function such that $f(t) = \pi - |t|$ for $t \in [-\pi; \pi]$.

a) Compute its Fourier coefficients.

b) Compute the sums of the series

$$\sum_{n=0}^{\infty} \frac{1}{(2n+1)^2}, \quad \sum_{n=1}^{\infty} \frac{1}{n^2}, \quad \sum_{n=0}^{\infty} \frac{1}{(2n+1)^4}, \quad \sum_{n=1}^{\infty} \frac{1}{n^4}.$$

Exercise 3.3. [p. 191] Let f be the 2π-periodic function defined by $f(t) = t(\pi - |t|)$, for $t \in [-\pi; \pi]$.

a) Compute its Fourier coefficients.

3.7. The uncertainty principle in communication theory.   133

b) Compute the sums of the series

$$\sum_{n=0}^{\infty}\frac{(-1)^n}{(2n+1)^3}, \quad \sum_{n=0}^{\infty}\frac{1}{(2n+1)^6}, \quad \sum_{n=1}^{\infty}\frac{1}{n^6}.$$

Exercise 3.4. [p. 193] Let a be a real number and let f be a 2π-periodic such that $f(t) = \cos(at)$ for $t \in]-\pi; \pi[$.
a) Compute its Fourier coefficients.
b) For $a \notin \mathbf{Z}$, prove that

$$\pi \cotan(\pi a) = \frac{1}{a} + \sum_{n=1}^{\infty}\frac{2a}{a^2 - n^2}.$$

c) Establish the formula

$$\sin(\pi a) = \pi a \prod_{n=1}^{\infty}\left(1 - \frac{a^2}{n^2}\right).$$

Exercise 3.5. [p. 195] Let p be a nonzero real number and let f be the 2π-periodic function such that $f(x) = e^{px}$ for $0 \leqslant x < 2\pi$.
a) Compute its complex Fourier coefficients.
b) Compute $\displaystyle\sum_{n=0}^{\infty}\frac{1}{p^2 + n^2}$.

Exercise 3.6. [p. 195] Let f, g be piecewise continuous periodic functions, with period $T > 0$. We define their convolution product $f * g$ by the formula

$$f * g(x) = \frac{1}{T}\int_0^T f(t)g(x - t)\,dt,$$

for any $x \in \mathbf{R}$.
a) Prove that $f * g$ is periodic, with period T.
b) Prove that it is continuous. (*First treat the case where f is a step function, introducing a primitive of g. Deduce the general case by approximation.*)
c) Prove that the Fourier coefficients of $f * g$ satisfy

$$c_n(f * g) = c_n(f)c_n(g),$$

for any $n \in \mathbf{Z}$.

Exercise 3.7. [p. 196] We propose another proof of the convergence of Fourier series, due to CHERNOFF (1980) and REDHEFFER (1984).
Let f be a continuous and T-periodic function from $\mathbf{R}$ to $\mathbf{C}$. Let $a \in \mathbf{R}$.
a) We assume that f is differentiable at a. Let F be the function defined by $F(t) = (f(t) - f(a))/(e^{2i\pi(t-a)/T} - 1)$ for $t \neq a \pmod{T}$. Prove that F can be

extended to a continuous and T-periodic function. Compute the Fourier coefficients of f in terms of those of F.

b) Under the assumptions of the preceding question, prove that the Fourier series of f converges to $f(a)$ at $t = a$. Can you propose weaker assumptions on f than continuity and differentiability at a that give rise to the same result?

c) Let δ be a real number such that $0 < \delta < T/2$, and let φ be the T-periodic function defined, for $t \in [a - T/2; a + T/2]$, by $\varphi(t) = -1$ if $a - \delta < t < a$, by $\varphi(t) = 1$ if $a < t < a+\delta$, and by $\varphi(t) = 0$ otherwise. Compute its Fourier transform. Justify that it converges to 0 at $t = a$.

d) We assume that f is piecewise $\mathscr{C}^1$ in a neighborhood of a. Prove that the Fourier series of f converges to $\frac{1}{2}(f(a^-) + f(a^+))$ at a. (*Introduce $f - \lambda\varphi$ for an appropriate complex number λ.*)

Exercise 3.8. [p. 198]   For $x \in \mathbf{R}$ and any integer $n \geqslant 1$, we set

$$S_n(x) = \sum_{k=1}^{n} \frac{\sin((2k-1)x)}{2k-1}.$$

a) Let f be the 2π-periodic function such that $f(0) = f(\pi) = 0$, $f(x) = 1$ for $0 < x < \pi$, and $f(x) = -1$ for $-\pi < x < 0$. Compute its Fourier transform. Conclude that for $x \in \,]0;\pi[$, we have $S_n(x) \to \pi/4$.

b) Compute the derivative of the function S_n. Deduce that the local extrema of S_n on $[0;\pi]$ are the real numbers $m_k = S_n(k\pi/2n)$, for all integers k such that $0 \leqslant k \leqslant 2n$, and that it is a local maximum if k is odd, a local minimum if k is even.

c) Prove that $m_1 \geqslant m_3 \geqslant \cdots \geqslant m_{2\lfloor n/2 \rfloor - 1}$. Deduce that $\sup(S_n) = m_1$.

d) Prove that

$$\lim_{n \to +\infty} (\sup(S_n)) = \frac{2}{\pi} \int_0^\pi \sin(t)/t \, dt.$$

In particular, one has $\lim_n(\sup(S_n)) > \sup(\lim_n(S_n))$. (*Gibbs phenomenon*)

Exercise 3.9. [p. 201]
Let $f: \mathbf{R} \to \mathbf{R}$ be the function defined by $f(x) = e^{-\pi x^2}$, and let g be its Fourier transform.

a) Using the formula $f'(x) = -2\pi x f(x)$, prove that g is a solution of the differential equation $g'(y) = -2\pi y g(y)$. Conclude that there exists a $c \in \mathbf{R}$ such that $g = cf$.

b) Prove that $c^2 = 1$, then that $c = 1$. Conclude that $\int_\mathbf{R} e^{-\pi x^2} \, dx = 1$.

c) For any real number $t > 0$, set

$$\theta(t) = \sum_{n=-\infty}^{\infty} \exp(-\pi t n^2).$$

Prove that $\theta(1/t) = \sqrt{t}\,\theta(t)$. (*Functional equation of the θ-function*) (*Apply the Poisson summation formula.*)

3.7. The uncertainty principle in communication theory.

Exercise 3.10. [p. 202]   Set $f(x) = e^{-|x|}$.
  a) Compute the Fourier transform of f.
  b) What can you deduce from the Fourier inversion formula at $x = 0$? Check this relation directly?
  c) What can you deduce from Plancherel's theorem?

Exercise 3.11. [p. 203]   Let f be the function from $\mathbf{R}$ to $\mathbf{R}$ given by $f(x) = \sup(0, 1 - |x|)$.
  a) Compute its Fourier transform.
  b) Deduce the relation
$$\int_{-\infty}^{\infty} \left(\operatorname{sinc}(\pi y)\right)^2 dy = 1.$$

Exercise 3.12. [p. 203]   a) Let f be the function from $\mathbf{R}$ to $\mathbf{R}$ given by $f(x) = 1$ for $x \in [-1; 1]$, and $f(x) = 0$ otherwise. Compute its Fourier transform.
  b) Prove that the integral
$$\int_0^{\infty} \left(\frac{\sin(t)}{t}\right)^2 dt$$
converges, and computes its value.
  c) Prove that the integral
$$\int_0^{\infty} \frac{\sin(t)}{t} dt$$
converges, and compute its value. (*Integrate from* 0 *to* T *and integrate by parts.*)

Notation

$\mathbf{P}(A)$, probability of an event A	9
$(X = a)$, event on which the random variable X is equal to a	11
$\mathbf{E}(X)$, expectation of a discrete random variable X	13
$\mathbf{V}(X)$, variance of a random variable X	16
$\mathbf{P}(A \mid B)$, conditional probability of an event A given an event B	17
$\mathbf{E}(X \mid B)$, conditional expectation of a random variable X given an event B	19
$\mathbf{E}(X \mid Y)$, conditional expectation of a random variable X given a random variable Y	20
$\mathbf{V}(X \mid Y)$, conditional variance of a random variable X given a random variable Y	20
$H(X)$, entropy of a discrete random variable X	25
$H_a(X)$, entropy base a of a random variable X	25
$H(X \mid Y)$, conditional entropy of X given Y	29
$D(p, q)$, divergence of the law q with respect to the law p	32
$I(X, Y)$, mutual information of a pair (X, Y) of random variables	34
$X \perp_Y Z$, the variables X and Z are conditionally independent with respect to Z	36
$I(X, Z \mid Y)$, conditional mutual information of X, Z with respect to Y	37
$H(X), \overline{H}(X), \underline{H}(X)$, entropy rate, upper and lower entropy rates, of the stochastic process X	38
$I(C)$, transmission capacity of a memoryless channel C	75
$X \sim_C Y$, the random variables X and Y are adapted to C	75
f_Φ, coding function associated with a code Φ	81
g_Φ, decoding function associated with code Φ	81
$\tau(\Phi)$, transmission rate of a code Φ	81
$\lambda_m(\Phi)$, probability of an error in the transmission of a block m using the code Φ	81
$\lambda_{\max}(\Phi)$, maximal probability of transmission error for the code Φ	82
$\lambda_{\mathrm{av}}(\Phi)$, average probability of transmission error for the code Φ	82
$a_n(f), b_n(f), c_n(f)$, Fourier coefficients of a periodic function f	100

$S_n(f)$, partial sums of the Fourier series of a periodic function f 101
$\widehat{f}$, Fourier transform of an integrable function   119
sinc, cardinal sine function   119

Chapter 4.
Solutions of the exercises

4.0. Some bits of probability theory

Exercise 0.1. [p. 20]   *a)* By definition of a uniform law on an n-element set, the common value p of the probabilities $\mathbf{P}(X = a)$ is $1/n$.

b) Let $m = \lfloor n/2 \rfloor$ be the integer part of $n/2$; one thus has $m = n/2$ if n is even and $m = (n-1)/2$ if n is odd. The even integers between 1 and n are the integers $2, 4, \ldots, 2m$, so that

$$\mathbf{P}(X \text{ is even}) = \mathbf{P}(X = 2) + \mathbf{P}(X = 4) + \cdots + \mathbf{P}(X = 2m) = m/n.$$

By the same argument, $\mathbf{P}(X \text{ is a multiple of } d)$ is equal to $\lfloor n/d \rfloor / n$.

c) One has

$$\mathbf{E}(X) = \mathbf{P}(X = 1) + 2\mathbf{P}(X = 2) + \cdots + n\mathbf{P}(X = n)$$
$$= (1 + 2 + \cdots + n)\frac{1}{n} = \frac{n(n+1)}{2} \times \frac{1}{n} = (n+1)/2.$$

d) Recall the formula $\mathbf{V}(X) = \mathbf{E}(X^2) - \mathbf{E}(X)^2$. Then

$$\mathbf{E}(X^2) = \mathbf{P}(X = 1) + 2^2 \mathbf{P}(X = 2) + \cdots + n^2 \mathbf{P}(X = n)$$
$$= (1^2 + 2^2 + \cdots + n^2)\frac{1}{n} = \frac{n(n+1)(2n+1)}{6} \times \frac{1}{n} = (n+1)(2n+1)/6.$$

Consequently,

$$\mathbf{V}(X) = \frac{(n+1)(2n+1)}{6} - \frac{(n+1)^2}{4}$$
$$= \frac{n+1}{12}(2(2n+1) - 3(n+1)) = \frac{n+1}{12}(n-1) = \frac{n^2-1}{12}.$$

Exercise 0.2. [p. 21]   *a*) Two reasonings seem legitimate:
- The first one consists in the prisoner saying he had two chances in three of being executed, but he now has just one in two — an optimistic stand;
- The second consists in not reasoning at all and saying that once the decisions are made, nothing has changed — a neutral stand.

Which of these two reasonings is correct?

Let X be the random variable that indicates the number (1, 2, or 3) of the pardoned prisoner, and let Y be the random variable that indicates the number of the prisoner given by the ward. Say our prisoner is number 1. If he hears that prisoner 2 will be executed, the information he possesses is that $X \neq 2$, and we have to compare $\mathbf{P}(X = 1)$ and $\mathbf{P}(X = 1 \mid Y = 2)$.

By definition of a conditional probability, we have

$$\mathbf{P}(X = 1 \mid Y = 2) = \frac{\mathbf{P}(X = 1 \text{ and } Y = 2)}{\mathbf{P}(Y = 2)}.$$

By Bayes's law, this is equal to

$$\frac{\mathbf{P}(Y = 2 \mid X = 1)\mathbf{P}(X = 1)}{(\mathbf{P}(Y = 2 \mid X = 1)\mathbf{P}(X = 1) + \mathbf{P}(Y = 2 \mid X = 2)\mathbf{P}(X = 2) + \mathbf{P}(Y = 2 \mid X = 3)\mathbf{P}(X = 3))},$$

or

$$\frac{\mathbf{P}(Y = 2 \mid X = 1)\mathbf{P}(X = 1)}{\mathbf{P}(Y = 2 \mid X = 1)\mathbf{P}(X = 1) + \mathbf{P}(X = 3)},$$

since $\mathbf{P}(Y = 2 \mid X = 2) = 0$ and $\mathbf{P}(Y = 2 \mid X = 3) = 1$. Indeed, the ward gives the number of a prisoner that shall be executed, and it has to be 2 if prisoner number 3 has been pardoned. We have to compare this expression to $\mathbf{P}(X = 1)$.

Let's choose some numerical values, assuming that the pardoned prisoner has been so by (uniform) random choice: then $\mathbf{P}(X = 1) = \mathbf{P}(X = 2) = \mathbf{P}(X = 3) = 1/3$. On the other hand, if prisoner number 1 is pardoned, let us imagine that the ward has randomly chosen his answer, that is, $\mathbf{P}(Y = 2 \mid X = 1) = 1/2$. Then,

$$\mathbf{P}(X = 1 \mid Y = 2) = \frac{(1/2) \times (1/3)}{(1/2) \times (1/3) + (1/3)} = \frac{1/6}{3/6} = \frac{1}{3}.$$

In this case, $\mathbf{P}(X = 1 \mid Y = 2) = \mathbf{P}(X = 1) = 1/3$, and nothing changed.

For the general case, let us write $a_i = \mathbf{P}(X = i)$ and $p = \mathbf{P}(Y = 2 \mid X = 1)$. Then

$$\mathbf{P}(X = 1 \mid Y = 2) - \mathbf{P}(X = 1) = \frac{pa_1}{pa_1 + a_3} - a_1 = \frac{a_1}{pa_1 + a_3}(p(1 - a_1) - a_3),$$

so that the conclusion, optimistic, pessimistic or neutral, depends on the values of a_1, a_2, a_3, and p. Let us study two extreme cases.

If $p = 0$, then the ward always indicates prisoner number 3 if he may, and then our prisoner may be worried: since the ward has indicated prisoner 2, he couldn't indicate prisoner 3 and prisoner 1 will be executed.

4.0. Some bits of probability theory.

On the other hand, when $p = 1$, the ward indicates prisoner 2 whenever he can; in this case, one has $p(1 - a_1) - a_3 = 1 - a_1 - a_3 = a_2 \geq 0$ and prisoner 1 may feel optimistic.

b) The candidate had one chance in three to choose the door that hides the car. We shall prove that the probability that the car lies behind the third door is now 2/3.

This is in fact the same problem as the first one, in a less dramatic disguise: the car corresponds to a pardon, and the open door to one of the unfortunate prisoners. One then has $\mathbf{P}(X = 1 \mid Y = 2) = 1/3$. Consequently,

$$\mathbf{P}(X = 3 \mid Y = 2) = 1 - \mathbf{P}(X = 1 \mid Y = 2) = 2/3,$$

and the candidate should open the third door!

Exercise 0.3. [p. 21]   *a)* The random variable X takes positive values; its expectation, possibly infinite, is given by the sum

$$\mathbf{E}(X) = \sum_{n=0}^{\infty} n \mathbf{P}(X = n) = \sum_{n=0}^{\infty} n e^{-p} p^n \frac{1}{n!}.$$

Writing $n/n! = 0$ for $n = 0$ and $n/n! = 1/(n-1)!$ for $n \geq 1$, we have

$$\mathbf{E}(X) = \sum_{n=1}^{\infty} \frac{1}{(n-1)!} p^n e^{-p} = \sum_{n=0}^{\infty} \frac{1}{n!} p^n e^{-p} \, p = p,$$

since $\sum_{n=0}^{\infty} (p^n/n!) = e^p$.

b) We have $\mathbf{E}(X^2) = \sum_{n=0}^{\infty} n^2 p^n e^{-p} \frac{1}{n!}$. Writing $n^2 = n(n-1) + n$, we get

$$\mathbf{E}(X^2) = \sum_{n=0}^{\infty} \frac{n(n-1)}{n!} p^n e^{-p} + \sum_{n=0}^{\infty} \frac{n}{n!} p^n e^{-p}$$

$$= \sum_{n=2}^{\infty} \frac{1}{(n-2)!} p^n e^{-p} + \sum_{n=1}^{\infty} \frac{1}{(n-1)!} p^n e^{-p}$$

$$= p^2 \sum_{n=0}^{\infty} \frac{1}{n!} p^n e^{-p} + p \sum_{n=0}^{\infty} \frac{1}{n!} p^n e^{-p}$$

$$= p^2 + p.$$

To compute the variance of X, we then write

$$\mathbf{V}(X) = \mathbf{E}(X^2) - \mathbf{E}(X)^2 = (p^2 + p) - p^2 = p.$$

c) We wish to compute

$$\mathbf{E}(X^k) = \sum_{n=0}^{\infty} n^k p^n e^{-p} \frac{1}{n!},$$

and we copy the method of the preceding questions. The polynomials

$$1 = \binom{n}{0},\ n = \binom{n}{1},\ \frac{n(n-1)}{2} = \binom{n}{2},\ \ldots,\ \frac{n(n-1)\ldots(n-k+1)}{k!} = \binom{n}{k-1}$$

have degrees $0, 1, \ldots, l$, hence they form a basis of the space of polynomials of degree $\leqslant k$. Consequently, there exist real numbers $c_0, \ldots, c_k$ such that for any n,

$$n^k = c_0 \binom{n}{0} + \cdots + c_k \binom{n}{k}.$$

Then

$$n^k \frac{1}{n!} = c_0 \frac{1}{n!} + c_1 \frac{1}{(n-1)!} + \cdots + c_k \frac{1}{(n-k)!},$$

and the same proof as in a) shows that

$$\mathbf{E}(X^k) = c_0 + c_1 p + \cdots + c_k p^k.$$

If we wish to obtained a "closed formula" for the coefficients c_i, we may observe that if we differentiate $e^p \mathbf{E}(X^k)$ with respect to p, the term $n^k p^n$ becomes $n^{k+1} p^{n-1}$. In other words, up to being allowed to differentiate term by term the sum of a summable family, and we will admit that it is legitimate here, we have

$$\mathbf{E}(X^{k+1}) = p e^{-p} \frac{d}{dp} \left(e^p \mathbf{E}(X^k) \right).$$

For $k = 0$, we have $X^0 = 1$, and we get

$$\mathbf{E}(X) = p e^{-p} \frac{d}{dp}(e^p) = p.$$

For $k = 1$, this gives

$$\mathbf{E}(X^2) = p e^{-p} \frac{d}{dp}(p e^p) = p e^{-p}(e^p + p e^p) = p(p+1),$$

and, for $k = 3$,

$$\mathbf{E}(X^3) = p e^{-p} \frac{d}{dp}(p(p+1)e^p) = p(2p + 1 + p(p+1)) = p(p^2 + 3p + 1).$$

By induction, we obtain a sequence $(L_k(p))_k$ of polynomials in p such that $\mathbf{E}(X^k) = L_k(p)$ for every p. These polynomials are determined by the condition $L_0(p) = 1$ and the induction formula

$$L_{k+1}(p) = p e^{-p} \frac{d}{dp}\left(L_k(p) e^p\right) = p L_k'(p) + p L_k(p).$$

By induction, we can see that L_k has degree k and leading coefficient 1.

4.0. Some bits of probability theory.

Exercise 0.4. [p. 21]   *a*) Since the discrete random variable takes positive real values, its expectation is given by the formula

$$E(X) = \sum_{n=1}^{\infty} n P(X = n) = \sum_{n=1}^{\infty} n q p^{n-1}.$$

On the other hand, we can differentiate termwise the equality $\frac{1}{1-p} = \sum_{n=0}^{\infty} p^n$ and obtain

$$\frac{1}{(1-p)^2} = \sum_{n=1}^{\infty} n p^{n-1}.$$

Consequently,

$$E(X) = q \times \frac{1}{(1-p)^2} = \frac{1}{q}.$$

If we do not make use of this termwise differentiation theorem, we may also write

$$\sum_{n=1}^{\infty} n q p^{n-1} = q \sum_{n=1}^{\infty} \sum_{m=1}^{n} p^{n-1} = q \sum_{m=1}^{\infty} \sum_{n=m}^{\infty} p^{n-1} = \sum_{m=1}^{\infty} p^{m-1} = \frac{1}{1-p} = \frac{1}{q},$$

as above.

b) The variance of X is given by $V(X) = E(X^2) - E(X)^2$, and we have

$$E(X^2) = \sum_{n=1}^{\infty} n^2 P(X = n) = \sum_{n=1}^{\infty} n^2 q p^{n-1}.$$

To compute the expectation of X^2, the differentiation method is the most efficient. From the equality

$$\sum_{n=1}^{\infty} n p^n = p \sum_{n=1}^{\infty} n p^{n-1} = \frac{p}{(1-p)^2},$$

we obtain

$$\sum_{n=1}^{\infty} n^2 p^{n-1} = \frac{1}{(1-p)^2} + 2 \frac{p}{(1-p)^3} = \frac{1+p}{(1-p)^3},$$

so that

$$E(X^2) = \frac{1+p}{q^2}.$$

Then

$$V(X) = \frac{1+p}{q^2} - \frac{1}{q^2} = \frac{p}{q^2}.$$

c) To compute the expectation of X^k, let us again use the differentiation method. By definition,

$$E(X^k) = q \sum_{n=1}^{\infty} n^k p^{n-1}.$$

If we differentiate $pq^{-1}\mathbf{E}(X^k)$ with respect to p, we obtain

$$\frac{d}{dp}\left(pq^{-1}\mathbf{E}(X^k)\right) = \frac{d}{dp}\left(\sum_{n=1}^{\infty} n^k p^n\right) = \sum_{n=0}^{\infty} n^{k+1} p^{n-1} = \frac{1}{q}\mathbf{E}(X^{k+1}),$$

so that

$$q^{-1}\mathbf{E}(X^{k+1}) = \frac{d}{dp}\left(pq^{-1}\mathbf{E}(X^k)\right).$$

For every integer k, let us set $f_k(p) = q^{-1}\mathbf{E}(X^k)$. These functions satisfy the induction relation $f_{k+1}(p) = (p f_k)'(p)$. For $k = 0$, we have $f_0(p) = q^{-1}\mathbf{E}(X^0) = 1/(1-p)$; we also have $f_1(p) = 1/(1-p)^2$ et $f_2(p) = (1+p)/(1-p)^3$.

Let us prove by induction that there exists a sequence $(S_k(p))$, where S_k is a monic polynomial of degree k, such that $f_k(p) = S_k(p)/(1-p)^{k+1}$. We just checked it for $k \leq 2$ with $S_0(p) = 1$, $S_1(p) = p$ and $S_2(p) = p + p^2$. Assume we have $f_k(p) = S_k(p)/(1-p)^{k+1}$, then

$$\begin{aligned}
f_{k+1}(p) &= \frac{d}{dp}\left(\frac{pS_k(p)}{(1-p)^{k+1}}\right) \\
&= \frac{S_k(p) + pS'_k(p)}{(1-p)^{k+1}} + (k+1)\frac{pS_k(p)}{(1-p)^{k+2}} \\
&= \frac{(1-p)S_k(p) + p(1-p)S'_k(p) + (k+1)pS_k(p)}{(1-p)^{k+2}},
\end{aligned}$$

which is a relation of the form required, with

$$S_{k+1}(p) = (k+1)pS_k(p) + p(1-p)S'_k(p) + (1-p)S_k(p).$$

If $S_k(p)$ is monic of degree p, then the degree of $S_{k+1}(p)$ is $\leq k+1$, and the coefficient of p^{k+1} is equal to $(k+1) - k = 1$, so that $S_{k+1}(p)$ is monic, of degree $k+1$, as desired.

Exercise 0.5. [p. 21]   a) Let A be the event "the chosen die is the unfair one" and let $\overline{A}$ be the complement event. One has $\mathbf{P}(A) = 1/3$ and $\mathbf{P}(\overline{A}) = 2/3$.

b) Let X be the event "the chosen die shows 1". One has $\mathbf{P}(X \mid A) = 2/3$ and $\mathbf{P}(X \mid \overline{A}) = 1/6$, so that

$$\mathbf{P}(X) = \mathbf{P}(A)\mathbf{P}(X \mid A) + \mathbf{P}(\overline{A})\mathbf{P}(X \mid \overline{A}) = \frac{1}{3} \times \frac{2}{3} + \frac{2}{3} \times \frac{1}{6} = \frac{1}{3}.$$

Applying Bayes's law, we then have

$$\mathbf{P}(A \mid X) = \frac{\mathbf{P}(X \mid A)\mathbf{P}(A)}{\mathbf{P}(X)} = \frac{(2/3) \times (1/3)}{1/3} = \frac{2}{3}.$$

Given this experiment, the probability that the chosen die is the unfair one is now 2/3!

4.0. Some bits of probability theory. 145

c) Let Y be the event "at the second throw, the die shows 1". Conditionally to A, this event is independent of X, but their conditional laws are identical: $P(Y \mid A) = 2/3$, $P(Y \mid \overline{A}) = 1/6$, $P(Y) = 1/3$.

By conditional independence given A, we then have

$$P(X \cap Y \mid A) = P(X \mid A)P(Y \mid A) = 4/9,$$

and

$$P(X \cap Y \mid \overline{A}) = P(X \mid \overline{A})P(Y \mid \overline{A}) = 1/36.$$

Finally,

$$P(X \cap Y) = P(A)P(X \cap Y \mid A) + P(\overline{A})P(X \cap Y \mid \overline{A})$$
$$= \frac{1}{3} \times \frac{4}{9} + \frac{2}{3} \times \frac{1}{36} = \frac{1}{6}.$$

Therefore,

$$P(A \mid X \cap Y) = \frac{P(X \cap Y \mid A)P(A)}{P(X \cap Y)} = \frac{(4/9) \times (1/3)}{1/6} = \frac{8}{9}.$$

Given these two experiments, the probability that the chosen die is unfair is now $8/9$.

Exercise 0.6. [p. 22] *a*) The random variable (X, Y) follows a uniform law: each of the 36 pairs (a, b), where $a, b \in \{1, \ldots, 36\}$, appears with probability $1/36$. Their sum Z belongs to $\{2; \ldots; 12\}$; the probability that Z equals a given integer s is thus equal to $n(s)/36$, where $n(s)$ if the number of pairs (a, b) such that $1 \leq a, b \leq 6$ and $a + b = s$.

We find $n(2) = 1$, $n(3) = 2, \ldots, n(7) = 6$, then $n(8) = 5, \ldots$, and finally $n(12) = 1$.

b) If $Z = 2$, then necessarily $X = 1$ so that $P(X = 1 \mid Z = 2) = 1$, hence $E(X \mid Z = 2) = 1$.

Conditionally given $Z = 3$, one has $X = 1$ or $X = 2$, each of the two issues having probability $1/2$, so that $E(X \mid Z = 3) = (1 + 2)/2 = 3/2$.

More generally, if $Z = a$, one has $\sup(a - 6, 1) \leq X \leq \inf(a - 1, 6)$, and given this event $(Z = a)$, each of the issues has the same probability. The conditional expectation is thus given by

$$E(X \mid Z = a) = \frac{1}{2}\bigl(\sup(a - 6, 1) + \inf(a - 1, 6)\bigr).$$

For $2 \leq a \leq 7$, one has $\sup(a - 6, 1) = 1$ and $\inf(a - 1, 6) = a - 1$, so that $E(X \mid Z = a) = \bigl(1 + (a - 1)\bigr)/2 = a/2$. For $7 \leq a \leq 12$, one has $\sup(a - 6, 1) = a - 6$ and $\inf(a - 1, 6) = 6$, so that $E(X \mid Z = a) = \bigl((a - 6) + 6\bigr)/2 = a/2$. Finally, we have

$$E(X \mid Z = a) = \frac{1}{2}a.$$

c) The random variable $\mathbf{E}(X \mid Z)$ is defined by

$$\mathbf{E}(X \mid Z)(\omega) = \mathbf{E}\big(X \mid Z = Z(\omega)\big)$$

for all $\omega \in \Omega$. Therefore,

$$\mathbf{E}(X \mid Z)(\omega) = Z(\omega)/2,$$

and $\mathbf{E}(X \mid Z) = Z/2$.

d) We can prove this relation in general. Since X and Y are independent, we have

$$\mathbf{P}(X = b \mid Z = a) = \frac{\mathbf{P}(X = b \text{ and } Y = a - b)}{\mathbf{P}(Z = a)} = \frac{\mathbf{P}(X = b)\mathbf{P}(Y = a - b)}{\mathbf{P}(Z = a)},$$

for all a, b such that $\mathbf{P}(Z = a) > 0$. Since X and Y have the same law, it follows by symmetry that

$$\mathbf{P}(X = b \mid Z = a) = \mathbf{P}(Y = b)\mathbf{P}(X = a - b)/\mathbf{P}(Z = a) = \mathbf{P}(Y = b \mid Z = a).$$

In particular, we have $\mathbf{E}(X \mid Z = a) = \mathbf{E}(Y \mid Z = a)$. By additivity of the conditional expectation, we thus have

$$2\,\mathbf{E}(X \mid Z = a) = \mathbf{E}(X \mid Z = a) + \mathbf{E}(Y \mid Z = a) = \mathbf{E}(X + Y \mid Z = a)$$
$$= \mathbf{E}(Z \mid Z = a) = a,$$

hence the relation $\mathbf{E}(X \mid Z = a) = a/2$, from which we deduce as above that $\mathbf{E}(X \mid Z) = Z/2$.

e) The random variables X and Y have 0 and 1 for possible values, and $\mathbf{P}(X = 1) = \mathbf{P}(Y = 1) = p$. Consequently, their product only has 0 and 1 for possible values. Since $XY = 1$ implies $X = Y = 1$, we have $\mathbf{P}(XY = 1) = p^2$.

Since $\mathbf{P}(X = 1 \mid XY = 1) = 1$, we have $\mathbf{E}(X \mid XY = 1) = 1$.

The event $(XY = 0)$ has probability $1 - p^2$, corresponding to the three other possibilities for the pair (X, Y), with respective probabilities

$$\mathbf{P}\big((X, Y) = (1, 0)\big) = p(1 - p),$$
$$\mathbf{P}\big((X, Y) = (0, 1)\big) = p(1 - p),$$
$$\mathbf{P}\big((X, Y) = (0, 0)\big) = (1 - p)^2.$$

Conditionally to the event $(XY = 1)$, their probabilities are thus

$$\frac{p(1-p)}{1-p^2} = \frac{p}{1+p}, \quad \frac{p}{1+p} \quad \text{and} \quad \frac{(1-p)^2}{1-p^2} = \frac{1-p}{1+p}.$$

Finally,

$$\mathbf{E}(X \mid XY = 0) = 1 \times \frac{p}{1+p} + 0 \times \frac{p}{1+p} + 0 \times \frac{1-p}{1+p} = \frac{p}{1+p}.$$

4.0. Some bits of probability theory.   147

If we wish to give a formula that is analogous to that of the preceding question, we may look for a function of a that is equal to $p/(1 + p)$ when $a = 0$ and 1 when $a = 1$. One of them is $a \mapsto (p + a)/(p + 1)$, hence the relation

$$\mathbf{E}(X \mid XY) = (p + XY)/(p + 1).$$

Exercise 0.7. [p. 22]   *a*) We shall start from the definition of the expectation as $\mathbf{E}(X) = \sum_x \mathbf{P}(X = x)x$, in which we insert the law of total probabilities: $\mathbf{P}(X = x) = \sum_y \mathbf{P}(X = x \text{ and } Y = y)$. We get

$$\begin{aligned}
\mathbf{E}(X) &= \sum_x \mathbf{P}(X = x)x \\
&= \sum_{x,y} \mathbf{P}(X = x \text{ and } Y = y)x \\
&= \sum_{\substack{y \\ \mathbf{P}(Y=y)>0}} \mathbf{P}(Y = y) \sum_x \mathbf{P}(X = x \mid Y = y)x \\
&= \sum_{\substack{y \\ \mathbf{P}(Y=y)>0}} \mathbf{P}(Y = y)\mathbf{E}(X \mid Y = y).
\end{aligned}$$

On the other hand, the restriction of the random variable $\mathbf{E}(X \mid Y)$ to the event $(Y = y)$ is constant: if $\mathbf{P}(Y = y) > 0$, its value is $\mathbf{E}(X \mid Y = y)$, and if $\mathbf{P}(Y) = 0$, it is zero. Consequently, the latter formula coincides with the definition of the expectation $\mathbf{E}(\mathbf{E}(X \mid Y))$, hence the result.

b) Recall that $V(X) = \mathbf{E}(X^2) - \mathbf{E}(X)^2$. On the other hand, we can write

$$\begin{aligned}
\mathbf{E}(V(X \mid Y)) &= \sum_y \mathbf{P}(Y = y)\mathbf{E}(X^2 \mid Y = y) - \sum_y \mathbf{P}(Y = y)\mathbf{E}(X \mid Y = y)^2 \\
&= \mathbf{E}(\mathbf{E}(X^2 \mid Y)) - \mathbf{E}(\mathbf{E}(X \mid Y)^2) \\
&= \mathbf{E}(X^2) - \mathbf{E}(\mathbf{E}(X \mid Y)^2),
\end{aligned}$$

by the first question. Similarly,

$$\begin{aligned}
V(\mathbf{E}(X \mid Y)) &= \mathbf{E}(\mathbf{E}(X \mid Y)^2) - \mathbf{E}(\mathbf{E}(X \mid Y))^2 \\
&= \mathbf{E}(\mathbf{E}(X \mid Y)^2) - \mathbf{E}(X)^2.
\end{aligned}$$

Then,

$$\begin{aligned}
\mathbf{E}(V(X \mid Y)) &+ V(\mathbf{E}(X \mid Y)) \\
&= \mathbf{E}(X^2) - \mathbf{E}(\mathbf{E}(X \mid Y)^2) + \mathbf{E}(\mathbf{E}(X \mid Y)^2) - \mathbf{E}(X)^2 \\
&= \mathbf{E}(X^2) - \mathbf{E}(X)^2 = V(X),
\end{aligned}$$

as was to be shown.

4.1. Entropy and mutual information

Exercise 1.1. [p. 49]   *a)* We interpret the hypothesis that the horses are of a comparable level as the fact that their position at the end of the race follow a uniform law. When it comes to the random variable X that gives Amy's gain, this leads to the following probabilities:

100	50	0	−10	−50	−100
1/8	2/8	1/8	1/8	2/8	1/8

Consequently,

$$E(X) = \frac{1}{8} \times 100 + \frac{2}{8} \times 50 + \frac{1}{8} \times 0 + \frac{1}{8} \times (-10) + \frac{2}{8} \times (-50) + \frac{1}{8} \times (-100)$$
$$= -\frac{10}{8} = -\frac{5}{4} = -1.25.$$

Since the law of X features 4 times the probability 1/8 and twice the probability 2/8, we get

$$H_2(X) = 4 \times \left(-\frac{1}{8} \log_2 \left(\frac{1}{8}\right)\right) + 2 \times \left(-\frac{2}{8} \log_2 \left(\frac{2}{8}\right)\right)$$
$$= \frac{1}{2} \log_2(8) + \frac{1}{2} \log_2(4) = \frac{1}{2} \times 3 + \frac{1}{2} \times 2 = \frac{5}{2}.$$

b) To compute the law of the random variable Y = |X|, we add up the probabilities that X = x and X = −x; this gives the following probabilities:

0	10	50	100
1/8	1/8	4/8	2/8

Consequently,

$$E(Y) = \frac{1}{8} \times 0 + \frac{1}{8} \times 10 + \frac{4}{8} \times 50 + \frac{2}{8} \times 100$$
$$= \frac{1}{8} (0 + 10 + 4 \times 50 + 2 \times 100)$$
$$= \frac{1}{8} (410) = \frac{205}{4} = 51.25.$$

Since the law of Y shows probability 1/8 twice, and once each the probabilities 2/8 and 4/8, we have

4.1. Entropy and mutual information.

$$H_2(Y) = 2 \times \left(-\frac{1}{8}\log_2\left(\frac{1}{8}\right)\right) + \left(-\frac{2}{8}\log_2\left(\frac{2}{8}\right)\right) + \left(-\frac{4}{8}\log_2\left(\frac{4}{8}\right)\right)$$
$$= \frac{1}{4}\log_2(8) + \frac{1}{4}\log_2(4) + \frac{1}{2}\log_2(2)$$
$$= \frac{1}{4} \times 3 + \frac{1}{4} \times 2 + \frac{1}{2} = \frac{7}{4} = 1.75.$$

In particular, we see that the entropy increased: $H_2(Y) \geq H_2(X)$.

Exercise 1.2. [p. 49]   a) (*With replacement*) The probability of picking a white ball is $w/(w+r)$, that of picking a red ball is $r/(w+r)$. Since we put the balls back in the urn, these probabilities are the same for the five balls and the random variables $X_1, \ldots, X_5$ have the same law: a Bernoulli law with parameter $p = w/(w+r)$. They are also independent, so that

$$H(X) = H(X_1) + H(X_2) + \cdots + H(X_5) = 5h(p).$$

b) (*Without replacement*) In this case, the random variables $Y_1, \ldots, Y_5$ are no longer independent, nor do they have the same law. For example, the probability that the first ball is white is $w/(w+r)$, the probability that the first two balls are white is

$$\frac{b}{b+r} \times \frac{b-1}{b+r-1} = \frac{b(b-1)}{(b+r)(b+r-1)},$$

but the probability that the second ball is white corresponds to the two possibilities (white, white) and (red, white), hence has probability

$$\frac{b(b-1)}{(b+r)(b+r-1)} + \frac{rb}{(b+r)(b+r-1)} = \frac{b(r+b-1)}{(b+r)(b+r-1)} = \frac{b}{b+r-1}.$$

Since

$$\mathbf{P}(Y_1 = \text{white and } Y_2 = \text{red}) = \frac{w(w-1)}{(w+r)(w+r-1)}$$
$$\neq \frac{w^2}{(w+r)^2} = \mathbf{P}(Y_1 = \text{white})\,\mathbf{P}(Y_2 = \text{white}),$$

the random variables Y_1 and Y_2 are not independent.

In this example, we see however that Y_1 and Y_2 have the same law! Let us prove that this holds in general. The following argument is simpler than a computational proof. Let us imagine that we pick n balls out of a bag containing N balls which are numbered from 1 to N; the sequences $(a_1, \ldots, a_n)$ of n distinct balls all have the same probability; since the numbers of the n balls that have been picked up can be switched arbitrarily, the laws of the a_i are all identical to the law of a_1. If we only recall the colors of the balls, we deduce that each Y_m follows a Bernoulli law with parameter $w/(w+r)$.

Consequently,

$$H(Y) = Y(Y_1, \ldots, Y_5)$$
$$= H(Y_1) + H(Y_2 \mid Y_1) + \cdots + H(Y_5 \mid Y_1, Y_2, Y_3, Y_4).$$

The entropy decreases by conditioning, so that $H(Y_2 \mid Y_1) \leq H(Y_2)$, etc. Finally,

$$H(Y) \leq H(Y_1) + \cdots + H(Y_5) = 5h(p) = H(X).$$

Exercise 1.3. [p. 49]   *a*) Since the law of X is uniform on the n-element set $\mathcal{X}$, one has the probabilities $p(x) = 1/n$ for all $x \in \mathcal{X}$. The entropy of X is then given by

$$H(X) = \sum_{i=1}^{n} -p(x_i) \log(p(x_i)) = n\left(-\frac{1}{n} \log\left(\frac{1}{n}\right)\right) = \log(n).$$

b) The function on $[0; 1]$ given by $x \mapsto x \log(x)$ (extended by 0 at 0) is continuous and it is twice differentiable on $]0; 1]$. Its first derivative is $x \mapsto \log(x) + 1$ and its second derivative is $x \mapsto 1/x$, hence it is strictly positive. This shows that the considered function is strictly convex and the desired inequality is exactly what strict convexity means: the average of the values of a function is smaller than the value at the average of the arguments.

c) Set $q_i = \mathbf{P}(Y = x_i)$ for each i and let us apply the preceding inequality: the average of the q_i is $(\mathbf{P}(Y = x_1) + \cdots + \mathbf{P}(Y = x_n))/n = 1/n$. Then

$$\frac{1}{n} \log\left(\frac{1}{n}\right) \leq \frac{1}{n} \sum_{i=1}^{n} \mathbf{P}(Y = x_i) \log\left(\mathbf{P}(Y = x_i)\right),$$

that is, $H(Y) \leq \log(n)$, and equality holds if and only if all q_i are equal, that is, if the law of Y is uniform.

d) By definition,

$$D(q, p) = \sum_{i=1}^{n} q(x_i) \log(q(x_i)/p(x_i))$$
$$= \sum_{i=1}^{n} q(x_i) \log(q(x_i)) - \sum_{i=1}^{n} q(x_i) \log(p(x_i))$$
$$= -H(Y) + \log(n) = H(X) - H(Y).$$

Since the divergence is always positive, we deduce that $H(Y) \leq H(X)$, and equality holds if and only if $D(q, p) = 0$, that is $q = p$, which means that the law of Y is uniform.

Exercise 1.4. [p. 50]   *a*) If $X = 0$, we got heads at the first toss of the coin; the probability that this happens is equal to x, hence $\mathbf{P}(X = 0) = x$.

The equality $X = 1$ means that we got tails at the first toss of the coin and head at the second; since coin tosses are independent, its probability is $\mathbf{P}(X = 1) = (1-x)x$.

4.1. Entropy and mutual information.

The equality $X = 2$ means that we obtained tails at the first two tosses, and heads at the third one, so that $\mathbf{P}(X = 2) = (1-x)^2 x$.

More generally, we have $\mathbf{P}(X = k) = (1-x)^k x$.

b) We have $\mathbf{E}(X) = \sum_{k=0}^{n} k \mathbf{P}(X = k) = \sum_{k=0}^{n} k(1-x)^k x$. Using the given formula, applied with $t = 1 - x$, this implies

$$\mathbf{E}(X) = x \frac{1-x}{\left(1-(1-x)\right)^2} = \frac{1-x}{x}.$$

c) By definition, we have

$$H(X) = -\sum_{k=0}^{\infty} \mathbf{P}(X = k) \log\left(\mathbf{P}(X = k)\right) = -\sum_{k=0}^{\infty} (1-x)^k x \log\left((1-x)^k x\right)$$

$$= \sum_{k=0}^{\infty} k(1-x)^k x \log(1-x) - \sum_{k=0}^{\infty} (1-x)^k x \times \log(x).$$

The first term of the latter expression is the expectation of X multiplied by $-\log(1-x)$, and the second term is $-\log(x)$; consequently,

$$H(X) = -\frac{1-x}{x} \log(1-x) - \log(x),$$

as was to be shown.

d) Set $h(t) = -(1-t)\log(1-t)/t - \log(t)$, so that $H(X) = h(x)$. The function h is differentiable on $]0; 1[$, and its derivative is given by

$$h'(t) = \frac{1}{t^2} \log(1-t) + \frac{1}{t} - \frac{1}{t} = \frac{1}{t^2} \log(1-t).$$

Since $h'(t) < 0$ for all t, the function h is strictly decreasing.

e) When t tends to 0, we have $\log(1-t) \sim -t$, hence $\log(1-t)/t \to -1$; on the other hand, $1 - t$ tends to 1 and $-\log(t)$ tends to $+\infty$, so that $\lim_{t \to 0} h(t) = +\infty$. In particular, there is a law of finite, arbitrarily large, entropy on $\mathbf{N}$, so no law maximizes entropy.

When t tends to 1, $1 - t$ tends to 0, hence $(1-t)\log(1-t)$ tends to 0 (by the indicated formula) hence $h(t)$ tends to 0.

f) By definition, we have

$$D(q, p) = \sum_{k=0}^{\infty} q(k) \log\left(q(k)/p(k)\right)$$

$$= \sum_{k=0}^{\infty} q(k) \log\left(q(k)\right) - \sum_{k=0}^{\infty} q(k) \log\left(p(k)\right).$$

The first term is equal to $H(Y)$. To compute the second, let us recall that $\mathbf{P}(X = k) = x^k(1-x)$, so that

$$\sum_{k=0}^{\infty} q(k) \log(p(k)) = \sum_{k=0}^{\infty} kq(k) \log(x) + \sum_{k=0}^{\infty} q(k) \log(1-x)$$
$$= \mathbf{E}(Y) \log(x) + \log(1-x),$$

and we recognize H(X). Consequently, $D(q, p) = H(X) - H(Y)$.

4.1. Entropy and mutual information. 153

g) Since the divergence is always positive, we conclude that $H(Y) \leq H(X)$: the entropy of a random variable with values in $\mathbf{N}$ is always smaller than or equal to the entropy of a geometric random variable with the same expectation. Equality only holds for $p = q$, that is, if Y is itself a geometric random variable.

The expectation of a geometric law with parameter x is given by $\mathbf{E}(X) = (1-x)/x$; the relation $\mathbf{E}(X) = m$ is then equivalent to $m = (1-x)/x = 1/x - 1$, hence $x = 1/(1+m)$. Then $1 - x = m/(1+m)$, so that

$$H(Y) \leq -m \log\left(m/(1+m)\right) - \log\left(1/(1+m)\right) = (m+1)\log(m+1) - m\log(m),$$

with equality if and only if Y is a geometric random variable with expectation m.

Exercise 1.5. [p. 50]   *a*) By definition of the conditional entropy, we have

$$H(f(X) \mid X) = \sum_{x \in \mathcal{X}} \mathbf{P}(X = x) H(f(X) \mid X = x),$$

where the sum is limited to those $x \in \mathcal{X}$ such that $\mathbf{P}(X = x) > 0$. Once restricted to the event $(X = x)$, the discrete random variable $f(X)$ becomes certain with value $f(x)$, so that $H(f(X) \mid X = x) = 0$. The equality $H(f(X) \mid X) = 0$ follows. Applying the chain rule for entropy, we deduce

$$H(X, f(X)) = H(X) + H(f(X) \mid X) = H(X).$$

b) Similarly, we have

$$H(X, f(X)) = H(f(X)) + H(X \mid f(X)) \geq H(f(X)),$$

since $H(X \mid f(X)) \geq 0$. Combining this inequality with the result of the preceding question we obtain the desired inequality, $H(X) \geq H(f(X))$.

c) Let us assume that f is injective; for $y \in \mathcal{Y}$ that belongs to the image of $\mathcal{X}$, let $g(y)$ be the unique preimage of y; otherwise, let us choose an arbitrary value for $g(y)$. This furnishes a map $g: \mathcal{Y} \to \mathcal{X}$ such that $g \circ f(x) = x$ for all $x \in \mathcal{X}$, hence $g(f(X)) = X$. By the preceding question, we have $H(f(X)) \geq H(g(f(X)) = H(X)$. Necessarily $H(X) = H(f(X))$.

d) The same argument as in the preceding question still works when one only assumes that $f|_{\mathcal{X}'}$ is injective. Indeed, let us denote by $g(y)$ the unique preimage in $\mathcal{X}'$ of an element $y \in f(\mathcal{X}')$, and let us choose $g(y)$ arbitrarily otherwise. Then $g \circ f(x) = x$ for every $x \in \mathcal{X}'$; since the event $X \in \mathcal{X}'$ has probability 1, the discrete random variables $g(f(X))$ and X are almost surely equal. They thus have the same law, hence the same entropy, so that $H(X) \geq H(f(X)) \geq H(g(f(X)) = H(X)$, hence again the equality $H(X) = H(f(X))$.

Conversely, let us assume that $H(X) = H(f(X))$. Rewriting the argument of question *c*), we observe that $H(X \mid f(X)) = 0$. By definition of the conditional entropy, we then have

$$0 = H(X \mid f(X))$$
$$= \sum_{y \in \mathcal{Y}} \mathbf{P}(f(X) = y) H(X \mid f(X) = y)$$
$$= \sum_{y \in \mathcal{Y}} \mathbf{P}(X \in f^{-1}(y)) H(X \mid X \in f^{-1}(y)),$$

where the sum is restricted to the set $\mathcal{Y}'$ of all these $y \in \mathcal{Y}$ such that $\mathbf{P}(X \in f^{-1}(y)) > 0$; in fact, one has $\mathcal{Y}' = f(\mathcal{Y})$. This implies that for $y \in \mathcal{Y}'$, the random variable $X \mid (X \in f^{-1}(y))$ conditioned to the event $(X \in f^{-1}(y))$ is certain. On the other hand, it takes every value of $\mathcal{X}' \cap f^{-1}(y)$ with a strictly positive probability; necessarily, $\mathcal{X}' \cap f^{-1}(y)$ has a single element, which means that $f|_{\mathcal{X}'}$ is injective.

e) Let us go back to the definition of the conditional entropy $H(Y \mid X)$: we have

$$H(Y \mid X) = \sum_{x \in \mathcal{X}'} \mathbf{P}(X = x) H(Y \mid X = x).$$

For every $x \in \mathcal{X}'$, the discrete random variable $Y \mid (X = x)$ conditioned by the event $(X = x)$ is therefore certain: it takes one value $y \in \mathcal{Y}$ with probability 1. If we denote this value by $g(x)$, we then obtain $\mathbf{P}(Y = g(x) \mid X = x) = 1$, hence

$$\mathbf{P}(X = x \text{ and } Y = g(x)) = \mathbf{P}(X = x),$$

so that

$$\mathbf{P}(X = x \text{ and } Y \ne g(x)) = 0,$$

since the event $(X = x)$ is the disjoint union of the two events $(X = x \text{ and } Y = g(x))$ and $(X = x \text{ and } Y \ne g(x))$. For $x \in \mathcal{X} - \mathcal{X}'$, let us choose an arbitrary element $g(x)$. The probability that $Y \ne g(X)$ is the sum of the probabilities $\mathbf{P}(X = x \text{ and } Y \ne g(x))$, so is zero. This implies that the random variables Y and $g(X)$ are almost surely equal.

Conversely, if $Y = g(X)$ (almost surely), we have $\mathbf{P}(Y = y \mid X = x) = 1$ if $y = g(x)$ and 0 otherwise, for every $x \in \mathcal{X}$ such that $\mathbf{P}(X = x) > 0$. For every such x, we thus have $H(Y \mid X = x) = 0$, hence $H(Y \mid X) = 0$.

Exercise 1.6. [p. 51]   *a*) For $y \in \{1, \ldots, 52\}$, we have

$$\mathbf{P}(Y = y) = \sum_{s \in \mathfrak{S}_{52}} \mathbf{P}(Y = b \text{ and } S = s)$$
$$= \sum_{s \in \mathfrak{S}_{52}} \mathbf{P}(X = s^{-1}(y) \text{ and } \mathbf{P}(S = s)) \quad \text{since } Y = S(X)$$
$$= \sum_{s \in \mathfrak{S}_{52}} \mathbf{P}(X = s^{-1}(y)) \mathbf{P}(S = s) \quad \text{by independence of } X \text{ and } S$$
$$= \frac{1}{\text{Card}(\mathfrak{S}_{52})} \sum_{s \in \mathfrak{S}_{52}} \mathbf{P}(X = s^{-1}(y)),$$

4.1. Entropy and mutual information.

because S is assumed to be uniform. When s runs among $\mathfrak{S}_{52}$, the element $x = s^{-1}(y)$ runs among $\mathcal{X}$, and each value is taken the same number of times, namely $\mathrm{Card}(\mathfrak{S}_{52})/\mathrm{Card}(\mathcal{X})$. Consequently,

$$\mathbf{P}(Y = y) = \frac{1}{\mathrm{Card}(\mathcal{X})} \sum_{x \in \mathcal{X}} \mathbf{P}(X = x) = \frac{1}{\mathrm{Card}(\mathcal{X})}.$$

This proves that the discrete random variable Y is uniform in $\mathcal{X}$. As we saw in exercise 1.3, its entropy is greater than the entropy of X.

b) Let us consider the random variable (S, X). By independence of X and S, we have $H(X, S) = H(X) + H(S)$. On the other hand, since the map $(x, s) \mapsto (s(x), s)$ is a permutation of the set $\mathcal{X} \times \mathfrak{S}_{52}$, one has $H(X, S) = H(S(X), S) = H(Y, S)$. By the chain rule, we have

$$H(Y, S) = H(S) + H(Y \mid S) \leq H(S) + H(Y).$$

Finally, we obtain $H(X) \leq H(Y)$.

Exercise 1.7. [p. 51]   a) By the law of total probabilities, we have

$$\mathbf{P}(X_1 = \mathrm{heads}) = \mathbf{P}(X_1 = \mathrm{heads} \mid Y = a)\mathbf{P}(Y = a) + \mathbf{P}(X_1 = \mathrm{tails} \mid Y = b)\mathbf{P}(Y = b)$$
$$= p \times \frac{1}{2} + q \times \frac{1}{2} = \frac{p + q}{2}.$$

The same computation holds for X_2, so that X_1 and X_2 have the same law.

b) By the same method, let us compute the probability that the coin fell with heads up twice:

$$\mathbf{P}(X_1 = \mathrm{heads} \text{ and } X_2 = \mathrm{heads})$$
$$= \mathbf{P}(X_1 = \mathrm{heads} \text{ and } X_2 = \mathrm{heads} \mid Y = a)\,\mathbf{P}(Y = a)$$
$$+ \mathbf{P}(X_1 = \mathrm{heads} \text{ and } X_2 = \mathrm{heads} \mid Y = b)\,\mathbf{P}(Y = b)$$
$$= p^2 \frac{1}{2} + q^2 \frac{1}{2}.$$

On the other hand,

$$\mathbf{P}(X_1 = \mathrm{heads})\,\mathbf{P}(X_2 = \mathrm{heads}) = (p + q)^2/4,$$

and

$$\frac{(p + q)^2}{4} - \frac{p^2 + q^2}{2} = \frac{2pq - p^2 - q^2}{4} = -\frac{(p - q)^2}{4}.$$

In other words, when $p \neq q$, the two random variables are not independent.

When $p = q$, the preceding computation shows that

$$\mathbf{P}(X_1 = \mathrm{heads} \text{ and } X_2 = \mathrm{heads}) = p^2 = \mathbf{P}(X_1 = \mathrm{heads})\,\mathbf{P}(X_2 = \mathrm{heads}).$$

And similarly for the three other cases, so that X_1 and X_2 are independent.

c) By definition,

$$I(X_1, X_2 \mid Y) = \mathbf{P}(Y = a)I(X_1, X_2 \mid Y = a) + \mathbf{P}(Y = b)I(X_1, X_2 \mid Y = b).$$

Once the first coin has been chosen (a choice which is witnessed by the random variable Y), the successive throws are independent: conditioned to the events $(Y = a)$ or $(Y = b)$, the random variables X_1 and X_2 are independent, so that the mutual informations $I(X_1, X_2 \mid Y = a)$ and $I(X_1, X_2 \mid Y = b)$ vanish. We thus have $I(X_1, X_2 \mid Y) = 0$ and $X_1 \perp_Y X_2$.

To compute $I(X_1, X_2)$, we list the possible probabilities

$X_1 \backslash X_2$	heads	tails
heads	$(p^2 + q^2)/2$	$(p(1-p) + q(1-q))/2$
tails	$(p(1-p) + q(1-q))/2$	$((1-p)^2 + (1-q)^2)/2$

as follows from the computation of the preceding question, and of the computations:

$$\mathbf{P}(X_1 = \text{heads and } X_2 = \text{tails}) = \mathbf{P}(X_1 = \text{heads}) - \mathbf{P}(X_1 = \text{heads and } X_2 = \text{heads})$$

$$= \frac{p+q}{2} - \frac{p^2 + q^2}{2} = \frac{p(1-p) + q(1-q)}{2},$$

a similar one for $\mathbf{P}(X_1 = \text{tails and } X_2 = \text{heads})$, while

$$\mathbf{P}(X_1 = \text{tails and } X_2 = \text{tails}) = \frac{(1-p)^2 + (1-q)^2}{2},$$

since exchanging the roles of heads and tails amounts to changing p into $1 - p$ and q into $1 - q$. Then the definition of mutual information implies

$$I(X_1, X_2) = \frac{p^2 + q^2}{2} \log\left(\frac{(p^2 + q^2)/2}{(p+q)^2/4}\right)$$

$$+ 2\frac{p(1-p) + q(1-q)}{2} \log\left(\frac{(p(1-p) + q(1-q))/2}{(p+q)(2-p-q)/4}\right)$$

$$+ \frac{(1-p)^2 + (1-q)^2}{2} \log\left(\frac{((1-p)^2 + (1-q)^2)/2}{(2-p+q)^2/4}\right).$$

It does not seem clear that this formula can be simplified.

Exercise 1.8. [p. 51]   *a)* The random variable Z can be written as $Z = \sigma(X, Y)$, where $\sigma: \mathbf{R}^2 \to \mathbf{R}$ is the sum map. Using exercise 1.5, we thus have $H(Z) \leq H(X, Y)$. On the other hand, $H(X, Y) \leq H(X) + H(Y)$. Consequently, $H(Z) \leq H(X) + H(Y)$.

4.1. Entropy and mutual information.

b) Let $x \in \mathbf{R}$ be such that $\mathbf{P}(X = x) > 0$; we have

$$H(Z \mid X = x) = \sum_{z \in \mathbf{R}} -\mathbf{P}(Z = z \mid X = x) \log \left(\mathbf{P}(Z = z \mid X = x)\right)$$
$$= \sum_{z \in \mathbf{R}} -\mathbf{P}(Y = z - x \mid X = x) \log \left(\mathbf{P}(Y = z - x \mid X = x)\right),$$

since, by definition of the random variable $Z = X + Y$, the events $(Z = z$ and $X = x)$ and $(Y = z - x$ and $X = x)$ are equal. By the change of summation index $y = z - x$, we then get

$$H(Z \mid X = x) = \sum_{y \in \mathbf{R}} -\mathbf{P}(Y = y \mid X = x) \log \left(\mathbf{P}(Y = y \mid X = x)\right)$$
$$= H(Y \mid X = x).$$

By definition of conditional entropy, it follows that

$$H(Z \mid X) = \sum_{x \in \mathbf{R}} \mathbf{P}(X = x) H(Z \mid X = x) = \sum_{x \in \mathbf{R}} \mathbf{P}(X = x) H(Y \mid X = x) = H(Y \mid X).$$

c) Let us assume that X and Y are independent. Since entropy decreases by conditioning, we have $H(Z) \geqslant H(Z \mid X)$. By the preceding question, we thus have $H(Z) \geqslant H(Y \mid X)$. Since X and Y are independent, we also have $H(Y \mid X) = H(Y)$, so that $H(Z) \geqslant H(Y)$. The other inequality, $H(Z) \geqslant H(X)$, holds similarly, by symmetry, so that

$$\sup \left(H(X), H(Y)\right) \leqslant H(Z).$$

d) When X and Y are independent, the preceding question shows that

$$\inf \left(H(X), H(Y)\right) \leqslant \sup \left(H(X), H(Y)\right) \leqslant H(Z).$$

We thus need to find an example where X and Y are dependent. In fact, let us take $Y = -X$, so that $Z = 0$ is certain, hence $H(Z) = 0$. However, if X is not certain, we have $H(X) = H(Y) > 0$, and the inequality $\inf \left(H(X), H(Y)\right) > H(Z)$ holds.

Exercise 1.9. [p. 51]   *a)* Since X and Y follow the same law, we have $H(X) = H(Y)$. Then

$$\rho(X; Y) = \frac{H(X) - H(Y \mid X)}{H(X)} = \frac{H(Y) - H(Y \mid X)}{H(X)} = \frac{I(X, Y)}{H(X)},$$

given the relation $H(Y) = H(Y \mid X) + I(X, Y)$. Using the formula $I(X, Y) + H(X, Y) = H(X) + H(Y)$, we see that that mutual information is symmetric in X and Y. Using once again that $H(X) = H(Y)$, we obtain

$$\rho(Y; X) = I(Y, X)/H(Y) = I(X, Y)/H(X) = \rho(X; Y).$$

158   4. Solutions of the exercises

b) Since the entropy is positive, we have $\rho(X;Y) \leq 1$. Since it decreases by conditioning, we also have $H(Y \mid X) \leq H(Y) = H(X)$, hence $\rho(X;Y) \geq 0$.

The case $\rho(X;Y) = 0$ means that $I(X,Y) = 0$, that is, that X and Y are independent.

The case $\rho(X;Y) = 1$ means that $H(Y \mid X) = 0$. Given exercise 1.5, there exists a function f such that $Y = f(X)$ (almost surely). Since we assumed that X and Y follow the same law, the function f cannot be arbitrary: one must have $\mathbf{P}(X = x) = \mathbf{P}(X = f(x))$ for all x. Conversely, for any function f for which this relation holds, the random variables X and $Y = f(X)$ follow the same law, and we have $\rho(X;Y) = 1$.

Exercise 1.10. [p. 51]   *a*) We have $H(X \mid Y) = H(X,Y) - H(Y)$ and $H(Y \mid X) = H(X,Y) - H(X)$, so that $\rho(X,Y) = 2H(X,Y) - H(X) - H(Y)$. Since $H(X) + H(Y) = H(X,Y) + I(X,Y)$, we also have the relation $\rho(X,Y) = H(X,Y) - I(X,Y)$.

b) Since conditional entropy is positive, we have $\rho(X,Y) = H(X \mid Y) + H(Y \mid X) \geq 0$, hence (i). The expression that defines $\rho(X,Y)$ is symmetric in X and Y, hence $\rho(X,Y) = \rho(Y,X)$, which proves (ii).

Given the relation $H(X \mid Y) = H(X,Y) - H(Y)$, we have

$$\rho(X,Y) + \rho(Y,Z) - \rho(X,Z)$$
$$= \big(2H(X,Y) - H(X) - H(Y)\big) + \big(2H(Y,Z) - H(Y) - H(Z)\big)$$
$$\quad - \big(2H(X,Z) - H(X) - H(Z)\big)$$
$$= 2\big(H(X,Y) + H(Y,Z) - H(X,Z) - H(Y)\big).$$

On the other hand, if we condition with respect to the variable Y, we get

$$H(X,Y) + H(Y,Z) = 2H(Y) + H(X \mid Y) + H(Z \mid Y)$$
$$= 2H(Y) + H(X,Z \mid Y) + I(X,Z \mid Y)$$
$$\geq 2H(Y) + H(X,Z \mid Y) = H(Y) + H(X,Z),$$

which proves relation (iii).

This computation also shows that the relation $\rho(X,Y) + \rho(Y,Z) = \rho(X,Z)$ is equivalent to $I(X,Z \mid Y) = 0$: the random variables X and Z are conditionally independent with respect to Y.

c) Since conditional entropy is positive, the equality $\rho(X,Y) = 0$ is equivalent to the vanishing of both $H(X \mid Y)$ and $H(Y \mid X)$. By exercise 1.3, the equality $H(X \mid Y) = 0$ means that there exists a function g such that $X = g(Y)$ (almost surely), while the equality $H(Y \mid X) = 0$ means that there exists a function f such that $Y = f(X)$ (almost surely). Consequently, f and g induce bijections from the set of possible values of X to the set of possible values of Y.

Exercise 1.11. [p. 52]   *a*) The upper entropy rate is the upper limit of the expressions $H(X_0, \ldots, X_{n-1})/n$ when n tends to infinity. For every n, we have

$$H(X_0, \ldots, X_{n-1}) \leq H(X_0) + \cdots + H(X_{n-1}) \leq n \log(a),$$

4.1. Entropy and mutual information.

since for each j, X_j takes its values in A and Card(A) $= a$, so that $H(X_j) \leq \log(a)$. Consequently, $H(X_0, \ldots, X_{n-1})/n \leq \log(a)$, and the upper entropy rate is at most $\log(a)$.

b) When the variables X_m are independent and follow the uniform law on A, this upper bound is an equality. However, this gives many ways in which this bound is non-optimal.

- Assume that the X_m are independent and follow the same law; in this case, we have $\overline{H}(X) = H(X_0)$, which only equals $\log(a)$ when the law of X_0 is uniform.
- Actually, we can relax the assumption that the X_m are independent: by the chain rule for entropy, we have

$$H(X_0, \ldots, X_{n-1}) = H(X_0) + H(X_1 \mid X_0) + \cdots + H(X_{n-1} \mid X_0, \ldots, X_{n-2})$$
$$\leq H(X_0) + H(X_1) + \cdots + H(X_{n-1}).$$

If the X_m all follow the same law, we conclude as above that $\overline{H}(X) = H(X_0)$.

- Take all X_m to be equal to the same uniform random variable on A. In this case, we have $H(X_0, \ldots, X_{n-1}) = H(X_0)$, hence $H(X_0, \ldots, X_{n-1})/n = H(X_0)/n$, which tends to 0, so that $\overline{H}(X) = 0$.

Exercise 1.12. [p. 52]   a) Conditioned to the event $(Y = a)$, the random variables X_k form a sequence of independent random variables that all follow the same law, a Bernoulli law with parameter p. Consequently,

$$H(X_1, \ldots, X_n \mid Y = a) = nH(X_1 \mid Y = a) = nh(p).$$

Conditioned to the event $(Y = b)$, we find similarly that

$$H(X_1, \ldots, X_n \mid Y = b) = nH(X_1 \mid Y = b) = nh(q).$$

It follows that

$$H(X_1, \ldots, X_n \mid Y) = \mathbf{P}(Y = a) H(X_1, \ldots, X_n \mid Y = a)$$
$$+ \mathbf{P}(Y = b) H(X_1, \ldots, X_n \mid Y = b)$$
$$= \frac{1}{2} nh(p) + \frac{1}{2} nh(q),$$

$$\lim_{n \to +\infty} \frac{1}{n} H(X_1, \ldots, X_n \mid Y) = \frac{1}{2}(h(p) + h(q)).$$

b) To estimate $H(X_1, \ldots, X_n)$, we introduce the random variable Y and write

$$H(X_1, \ldots, X_n) \leq H(X_1, \ldots X_n, Y)$$
$$= H(Y) + H(X_1, \ldots, X_n \mid Y)$$
$$= \log(2) + n\frac{1}{2}(h(p) + h(q)).$$

Dividing by n and making n grow to infinity, we obtain the upper bound

$$\overline{\mathrm{H}}(X) \leqslant \frac{1}{2}(h(p) + h(q)).$$

On the other hand, we also have

$$\mathrm{H}(X_1, \ldots, X_n, Y) = \mathrm{H}(X_1, \ldots, X_n) + \mathrm{H}(Y \mid X_1, \ldots, X_n)$$
$$\leqslant \mathrm{H}(X_1, \ldots, X_n) + \log(2),$$

since Y takes its values in $\{a, b\}$. Therefore,

$$\mathrm{H}(X_1, \ldots, X_n) \geqslant \mathrm{H}(X_1, \ldots X_n, Y) - \log(2)$$
$$= \mathrm{H}(Y) - \log(2) + \mathrm{H}(X_1, \ldots, X_n \mid Y)$$
$$= \mathrm{H}(Y) - \log(2) + n\frac{1}{2}(h(p) + h(q)).$$

Again, we divide by n and make n grow to infinity; this implies the lower bound:

$$\underline{\mathrm{H}}(X) \geqslant \frac{1}{2}(h(p) + h(q)).$$

Finally, the entropy rate of the random process X is equal to $\mathrm{H}(X) = \frac{1}{2}(h(p) + h(q))$.

c) When p and q are quite different, say when p is close to 0 and q is close to 1, intuition suggests that X is *not* a Markov chain. Assume for example that we first chose the coin a; in this case, the tosses X_n will mostly show heads, while if we had chosen the coin b, they will show tails: in addition to the knowledge of X_n, the knowledge of the tosses $(X_1, \ldots, X_{n-1})$ gives us some information about the chosen coin, hence about the coin X_{n+1}.

More precisely, writing 0/1 for tails/heads, we will compute

$$\mathbf{P}(X_{n+1} = 1 \mid X_m = \cdots = X_n = 1),$$

for $m = 1$ and for $m = n$.

By definition,

$$\mathbf{P}(X_{n+1} = 1 \mid X_m = \cdots = X_n = 1) = \frac{\mathbf{P}(X_m = \cdots = X_{n+1} = 1)}{\mathbf{P}(X_m = \cdots = X_n = 1)}.$$

We compute the numerator and denominator by using the law of total probabilities:

$$\mathbf{P}(X_m = \cdots = X_n = 1) = \mathbf{P}(Y = a)\mathbf{P}(X_m = \cdots = X_n = 1 \mid Y = a)$$
$$+ \mathbf{P}(Y = b)\mathbf{P}(X_m = \cdots = X_n = 1 \mid Y = b)$$
$$= \frac{1}{2}p^{n+1-m} + \frac{1}{2}q^{n+1-m}.$$

Similarly,

$$\mathbf{P}(X_m = \cdots = X_{n+1} = 1) = \frac{1}{2}p^{n+2-m} + \frac{1}{2}q^{n+2-m}.$$

4.1. Entropy and mutual information.

Consequently,

$$P(X_{n+1} = 1 \mid X_m = \cdots = X_n = 1) = \frac{p^{n+2-m} + q^{n+2-m}}{p^{n+1-m} + q^{n+1-m}}.$$

For $p = q$, we get p; in this case, the stochastic process X is markovian and homogeneous: from the probabilistic point of view, we cannot detect any difference between the two coins. On the other hand, let us assume that $p \neq q$, say $q < p$. When $m = 1$ and $n \to +\infty$, we obtain

$$P(X_{n+1} = 1 \mid X_1 = \cdots = X_n = 1) \to p,$$

while when $m = n$, we have

$$P(X_{n+1} = 1 \mid X_n = 1) = \frac{p^2 + q^2}{p + q} = p - \frac{(p-q)q}{p+q} < p,$$

which proves that the stochastic process X is not markovian.

Exercise 1.13. [p. 52]   *a)* One has $\mu_n = \mu_0 P^n$.

b) If $(p, q) = (1, 1)$, then $P = \begin{pmatrix} 0 & 1 \\ 1 & 0 \end{pmatrix}$, so that $P^n = I_2$ for even n, and $P^n = P$ for odd n. Consequently, $(u_n, v_n) = (u_0, v_0)$ for even n, and $(u_n, v_n) = (v_0, u_0)$ for odd n. Unless $u_0 = v_0 = 1/2$, the sequence (μ_n) does not have a limit.

If $(p, q) = (0, 0)$, then $P = I_2$, every law is stationary and the sequence (μ_n) is constant.

c) We have

$$P = \begin{pmatrix} 1-p & p \\ q & 1-q \end{pmatrix}.$$

Its trace is $2 - p - q$ and its determinant is $(1-p)(1-q) - pq = 1 - p - q$, so that its characteristic polynomial is given by

$$T^2 - (2 - p - q)T + (1 - p - q) = (T - 1)(T - 1 + p + q).$$

The two eigenvalues of P are 1 and $1 - p - q \in \,]-1\,;1[$; they are simple.

d) The column-vector $U = \begin{pmatrix} 1 \\ 1 \end{pmatrix}$ is a "right-eigenvector" for the eigenvalue 1: $PU = U$. Let us find an eigenvector V for the eigenvalue $\lambda = 1 - p - q$. If its coordinates are (x, y), we get the system

$$(1-p)x + py = (1 - p - q)x, \quad qx + (1-q)y = (1 - p - q)y,$$

hence $qx + py = 0$, so that we may take $V = \begin{pmatrix} -p \\ q \end{pmatrix}$. Let then $Q = (UV) = \begin{pmatrix} 1 & -p \\ 1 & q \end{pmatrix}$. Its inverse is the matrix

$$Q^{-1} = \frac{1}{p+q} \begin{pmatrix} q & p \\ -1 & 1 \end{pmatrix}.$$

By construction, we have $PQ = DQ$, where $D = \begin{pmatrix} 1 & 0 \\ 0 & 1-p-q \end{pmatrix}$. Therefore,

$$P^n = QD^nQ^{-1} = \frac{1}{p+q}\begin{pmatrix} 1 & -p \\ 1 & q \end{pmatrix}\begin{pmatrix} 1 & 0 \\ 0 & (1-p-q)^n \end{pmatrix}\begin{pmatrix} q & p \\ -1 & 1 \end{pmatrix}$$

$$= \frac{1}{p+q}\begin{pmatrix} 1 & -(1-p-q)^n p \\ 1 & (1-p-q)^n q \end{pmatrix}\begin{pmatrix} q & p \\ -1 & 1 \end{pmatrix}$$

$$= \frac{1}{p+q}\begin{pmatrix} q+(1-p-q)^n p & p-(1-p-q)^n p \\ q-(1-p-q)^n q & p+(1-p-q)^n q \end{pmatrix}.$$

e) By assumption, $|1-p-q| < 1$. It follows that when $n \to +\infty$, P^n tends to the matrix

$$\frac{1}{p+q}\begin{pmatrix} q & p \\ q & p \end{pmatrix}.$$

Since $\mu_n = \mu_0 P^n$, we conclude that (μ_n) converges to

$$\frac{1}{p+q}(u_0 \ v_0)\begin{pmatrix} q & p \\ q & p \end{pmatrix} = (u_0+v_0)\left(\frac{q}{p+q} \ \frac{p}{p+q}\right)$$

$$= \left(\frac{q}{p+q} \ \frac{p}{p+q}\right),$$

using that $u_0 + v_0 = 1$. In particular, we observe that (μ_n) converges to the stationary law.

Let us also observe that we have some control on the rate of convergence: it is governed by the term $(1-p-q)^n$, so that the convergence is exponential, and is faster the further we are from the two extreme cases $(p,q) = (0,0)$ and $(p,q) = (1,1)$.

Exercise 1.14. [p. 52]   a) Since two consecutive variables are independent, we have $\mathbf{P}(Y_n = 0) = (1-p)^2$, $\mathbf{P}(Y_n = 1) = 1 - p^2 - (1-p)^2 = 2(1-p)p$ and $\mathbf{P}(Y_n = 2) = p^2$. Consequently,

$$H(Y_n) = -2p^2 \log(p) - 2(1-p)^2 \log(1-p) - 2p(1-p)\log\big((1-p)p\big)$$
$$= -\big(2p^2 + 2p(1-p)\big)\log(p) - \big(2(1-p)^2 + 2(1-p)p\big)\log(1-p)$$
$$= -2p\log(p) - 2(1-p)\log(1-p) = 2h(p).$$

b) To compute $H(Y_n \mid Y_{n_1})$, we compute the various conditional probabilities.

Assume $Y_{n-1} = 0$; this means that $X_{n-1} = X_{n-2} = 0$, so that conditioned to this event, $Y_n = X_n$ follows a Bernoulli law with parameter p, hence

$$H(Y_n \mid Y_{n-1} = 0) = h(p).$$

Assume $Y_{n-1} = 1$; there are two possibilities, each equiprobable: either $X_{n-1} = 1$ and $X_{n-2} = 0$, or $X_{n-2} = 1$ and $X_{n-1} = 0$. In the first case, $Y_n = 1 + X_n$ takes the value 1 with probability $1-p$, and the value 2 with probability p; in the second case, $Y_n = X_n$ takes the value 1 with probability p and 0 with probability $1-p$. Consequently,

4.1. Entropy and mutual information.

$$P(Y_n = 0 \mid Y_{n-1} = 1) = \frac{1}{2}(1-p),$$

$$P(Y_n = 1 \mid Y_{n-1} = 1) = \frac{1}{2}(p+1-p) = \frac{1}{2},$$

$$P(Y_n = 2 \mid Y_{n-1} = 1) = \frac{1}{2}p,$$

hence

$$H(Y_n \mid Y_{n-1} = 1) = -\frac{1}{2}(1-p)\log\left(\frac{1}{2}(1-p)\right) - \frac{1}{2}\log\left(\frac{1}{2}\right) - \frac{1}{2}p\log\left(\frac{1}{2}\right)$$
$$= \log(2) + \frac{1}{2}h(p).$$

Let us finally assume $Y_{n-1} = 2$, so that $X_{n-1} = X_{n-2} = 1$; in this case, $Y_n = 1 + X_n$ takes the value 2 with probability p and the value 1 with probability $1-p$, so that $H(Y_n \mid Y_{n_1} = 2) = h(p)$.

To conclude, we have

$$H(Y_n \mid Y_{n-1}) = (1-p)^2 h(p) + 2p(1-p)\left(\log(2) + \frac{1}{2}h(p)\right) + p^2 h(p)$$
$$= \left(1 - 2p + p^2 + p(1-p) + p^2\right)h(p) + 2p(1-p)\log(2)$$
$$= (1 - p + p^2)h(p) + 2p(1-p)\log(2).$$

c) Let us prove that, unless $p = 1/2$, the random process (Y_n) is not markovian. For this, let us consider the probability $P(Y_3 = 1 \mid Y_1 = 0, Y_2 = 1)$. The conditioning event, $(Y_1 = 0, Y_2 = 1)$, is the intersection of the events $(X_0 = 0)$, $(X_1 = 0)$ and $(X_2 = 1)$, so that on restriction to it, we have $Y_3 = X_2 + X_3$. Consequently,

$$P(Y_3 = 1 \mid Y_1 = 0, Y_2 = 1) = P(X_3 = 0 \mid X_0 = 0, X_1 = 0, X_2 = 1) = 1 - p.$$

On the other hand, the event $(Y_2 = 1)$ is the union of two disjoint events, $((X_1, X_2) = (0, 1))$, and $((X_1, X_2) = (1, 0))$, each having probability $(1-p)p$, so that $P(Y_2 = 1) = 2(1-p)p$. On the other hand, the event $(Y_2 = Y_3 = 1)$ is the union of two disjoint events, $((X_1, X_2, X_3) = (0, 1, 0))$ and $((X_1, X_2, X_3) = (1, 0, 1))$, with respective probabilities $(1-p)^2 p$ and $(1-p)p^2$, so that

$$P(Y_2 = Y_3 = 1) = (1-p)^2 p + (1-p)p^2 = (1-p)p.$$

Finally,

$$P(Y_3 = 1 \mid Y_2 = 1) = \frac{(1-p)p}{2(1-p)p} = \frac{1}{2} \neq 1 - p = P(Y_3 = 1 \mid Y_1 = Y_2 = 1).$$

d) We have

$$H(Y_1, \ldots, Y_n) \leqslant H(X_0, Y_1, \ldots, Y_n) \leqslant H(X_0) + H(Y_1, \ldots, Y_n).$$

However, the knowledge of $(X_0, Y_1, \ldots, Y_n)$ is equivalent to that of $(X_0, \ldots, X_n)$, since $X_1 = Y_1 - X_0$, $X_2 = Y_2 - X_1$, etc. Therefore,

$$H(X_0, Y_1, \ldots, Y_n) = H(X_0, X_1, \ldots, X_n) = (n+1)h(p),$$

since the sequence (X_k) is independent and the random variables X_k all have entropy $h(p)$. This implies the inequality

$$nh(p) \leq H(Y_1, \ldots, Y_n) \leq (n+1)h(p),$$

from which we conclude that the entropy rate of the random process (Y_k) is equal to $h(p)$.

Exercise 1.15. [p. 52]   *a)* Going back to the definition, we have

$$D\big((X, X'), (Y, Y')\big) - D(X, Y)$$

$$= \sum_{a,b} P(X = a \text{ and } X' = b) \log \frac{P(X = a \text{ and } X' = b)}{P(Y = a \text{ and } Y' = b)}$$

$$- \sum_{a} P(X = a) \log \frac{P(X = a)}{P(Y = a)}$$

$$= \sum_{a,b} P(X = a \text{ and } X' = b) \log \frac{P(X = a \text{ and } X' = b)}{P(Y = a \text{ and } Y' = b)}$$

$$- \sum_{a} P(X = a \text{ and } X' = b) \log \frac{P(X = a)}{P(Y = a)}$$

$$= \sum_{a,b} P(X = a \text{ and } X' = b) \log \frac{P(X = a \text{ and } X' = b)/P(X = a)}{P(Y = a \text{ and } Y' = b)/P(Y = a)}$$

$$= \sum_{a} P(X = a) \sum_{b} P(X' = b \mid X = a) \log \frac{P(X' = b \mid X = a)}{P(Y' = b \mid Y = a)}.$$

By definition, for each fixed a, the expression

$$\sum_{b} P(X' = b \mid X = a) \log \frac{P(X' = b \mid X = a)}{P(Y' = b \mid Y = a)}$$

is the divergence of the law of Y' conditioned to the event $(Y = a)$ with respect to the law of X' conditioned to that event. This implies the required relation.

b) To simplify the notation, we write $p = p_n$, $q = q_n$, $p' = p_{n+1}$ and $q' = q_{n+1}$, as well as $X = X_n$, $Y = Y_n$ and $X' = X_{n+1}$, $Y' = Y_{n+1}$. We have to prove $D(p', q') \leq D(p, q)$.

Let us first apply the relation of the first question. For every a, the law of X' given $(X = a)$ coincides with the law of Y' given $(Y = a)$, so that $D\big((X, X'), (Y, Y')\big) = D(X, Y)$. On the other hand, we may apply the second part of the preceding question while exchanging the roles of X, Y and X', Y'; this gives the inequality

4.1. Entropy and mutual information.   165

$$D\big((X,X'),(Y,Y')\big) \geqslant D(X',Y').$$

Finally, we have

$$D(X,Y) = D\big((X,X'),(Y,Y')\big) \geqslant D(X',Y'),$$

as requested.

c) Since the Markov chain (Y_n) is stationary, the sequence (q_n) is constant. Since the Markov chain (X_n) is primitive, the sequence (p_n) converges to the unique stationary law, which is equal to q_0. Consequently, $D(p_n, q_n) = D(p_n, q_0)$ converges to 0.

d) Since the law q_0 is uniform, we have

$$\begin{aligned} D(p_n, q_n) &= D(p_n, q_0) \\ &= \sum_a p_n(a) \log \frac{p_n(a)}{1/\mathrm{Card}(A)} \\ &= \log\big(\mathrm{Card}(A)\big) - H(X_n), \end{aligned}$$

hence $H(X_n) = \log\big(\mathrm{Card}(A)\big) - D(p_n, q_n)$. This proves that the sequence $(H(X_n))$ is increasing and converges to $\log\big(\mathrm{Card}(A)\big)$.

Remark. — This is a variant of the second principle of thermodynamics: a gas evolves to equilibrium and its entropy increases. When X, Y are random variables with laws p, q, the divergence $D(p, q)$ is sometimes called the *relative entropy* of X and Y. Despite the close terminology, entropy and relative entropy behave differently.

Exercise 1.16. [p. 53]   a) This is nothing but the data processing inequality from corollary 1.3.12, as well as its characterization of the equality case.

b) To prove that f is a sufficient statistic for T, we check the conditional independence of T and X with respect to $f(X)$, that is to say, we establish the equalities

$$\mathbf{P}(T = t \mid X = a) = \mathbf{P}\big(T = t \mid f(X) = f(a)\big),$$

for all $a \in \{0,1\}^n$ and all $t \in \Theta$. This can be rewritten

$$\mathbf{P}(T = t \text{ and } X = a) \, \mathbf{P}\big(f(X) = f(a)\big) = \mathbf{P}\big(T = t \text{ and } f(X) = f(a)\big) \, \mathbf{P}(X = a),$$

hence

$$\frac{\mathbf{P}(f(X) = f(a) \mid T = t)}{\mathbf{P}(X = a \mid T = t)} = \frac{\mathbf{P}(f(X) = f(a))}{\mathbf{P}(X = a)}.$$

The Bernoulli assumption on the X_k implies that we have

$$\mathbf{P}(X = a \mid T = t) = t^m (1-t)^{n-m},$$

so that

$$\mathbf{P}(X = a \text{ and } T = t) = t^m (1-t)^{n-m} \mathbf{P}(T = t).$$

Since there are $\binom{n}{m}$ elements $a \in \{0,1\}^n$ for which $f(a) = m$, we obtain

$$\mathbf{P}(f(X) = f(a) \text{ and } T = t) = t^m(1-t)^{n-m}\mathbf{P}(T=t)/\binom{n}{m}.$$

Consequently,
$$\frac{\mathbf{P}(f(X) = f(a) \mid T = t)}{\mathbf{P}(X = a \mid T = t)} = 1/\binom{n}{m}.$$

The important point is that this expression does not depend on t. Then,
$$\mathbf{P}(f(X) = a) = \sum_t \mathbf{P}(f(X) = a \mid T = t)\mathbf{P}(T = t)$$
$$= \frac{1}{\binom{n}{m}} \sum_t \mathbf{P}(X = a \mid T = t)\mathbf{P}(T = t) = \frac{1}{\binom{n}{m}} \mathbf{P}(X = a).$$

c) By definition, we have $\theta^*(X) = \mathbf{E}(\theta(X) \mid f(X))$. Using the first question of exercise 0.7, this implies
$$\mathbf{E}(\theta^*(X)) = \mathbf{E}\big(\mathbf{E}(\theta(X) \mid f(X))\big) = \mathbf{E}(\theta(X)).$$

It is also possible to redo the argument: for every value k of f, the estimator $\theta^*(X)$ is constant conditionally to the event $(f(X) = k)$, with value $\mathbf{E}(\theta(X) \mid f(X) = k)$; it follows that
$$\mathbf{E}(\theta^*(X) \mid f(X) = k) = \mathbf{E}(\theta(X) \mid f(X) = k),$$
hence the desired equality of expectations.

On the other hand, applying the second question of exercise 0.7 to the pairs $(\theta(X) - T, f(X))$ and $(\theta^*(X) - T, f(X))$, we obtain
$$\mathbf{V}(\theta(X) - T) - \mathbf{V}(\theta^*(X))$$
$$= \mathbf{V}\big(\mathbf{E}(\theta(X) - T \mid f(X))\big) - \mathbf{V}\big(\mathbf{E}(\theta^*(X) - T \mid f(X))\big)$$
$$+ \mathbf{E}\big(\mathbf{V}(\theta(X) - T \mid f(X))\big) - \mathbf{E}\big(\mathbf{V}(\theta^*(X) - T \mid f(X))\big).$$

The two random values $\mathbf{E}(\theta(X) - T \mid f(X))$ and $\mathbf{E}(\theta^*(X) - T \mid f(X))$ are equal to $\theta^*(X) - \mathbf{E}(T \mid f(X))$, by definition of $\theta^*(X)$, so that the first two terms cancel each other. The two other are expectations; let us first consider a given value k for $f(X)$ and let us write
$$\mathbf{V}(\theta(X) - T \mid f(X) = k) - \mathbf{V}(\theta^*(X) - T \mid f(X) = k)$$
$$= \mathbf{E}\big((\theta(X) - T)^2 \mid f(X) = k\big) - \mathbf{E}\big((\theta^*(X) - T)^2 \mid f(X) = k\big),$$
since, restricting to the event $(f(X) = k)$, we have
$$\mathbf{E}(\theta(X) - T \mid f(X) = k) = \theta^*(X) - \mathbf{E}(T \mid f(X) = k)$$
$$= \mathbf{E}(\theta^*(X) - T \mid f(X) = k).$$

4.1. Entropy and mutual information.

Expanding the squares, this gives

$$\mathbf{E}\big(\theta(X)^2 - \theta^*(X)^2 \mid f(X) = k\big) + 2\mathbf{E}\big(T \times (\theta^*(X) - \theta(X)) \mid f(X) = k\big).$$

The last term vanishes, since f is a sufficient statistic for T: conditionally to the event $(f(X) = k)$, the random variables X and T are independent, so that

$$\mathbf{E}\big(T \times (\theta^*(X) - \theta(X)) \mid f(X) = k\big)$$
$$= \mathbf{E}(T \mid f(X) = k)\mathbf{E}\big(\theta^*(X) - \theta(X) \mid f(X) = k\big) = 0,$$

using the definition of $\theta^*(X)$. Moreover,

$$\mathbf{E}\big(\theta(X)^2 - \theta^*(X)^2 \mid f(X) = k\big) = \mathbf{V}\big(\theta(X) \mid f(X) = k\big),$$

so that

$$\mathbf{E}\big(\mathbf{V}(\theta(X) - T \mid f(X))\big) - \mathbf{E}\big(\mathbf{V}(\theta^*(X) - T \mid f(X))\big) = \mathbf{V}\big(\theta(X) \mid f(X)\big).$$

This implies the desired formula.

Exercise 1.17. [p. 53] *a*) The three states represent the action of the student, "b" for book, "c" for coffee and "a" when she goes into the open air. Let X_n be its state at "time" n. According to our little story, when the student reads her book, she may go for a coffee with probability p: we thus have $\mathbf{P}(X_{n+1} = c \mid X_n = b) = p$, and $\mathbf{P}(X_{n+1} = b \mid X_n = b) = 1 - p$. When she has finished her coffee, she goes back to the library with probability $1 - q$: $\mathbf{P}(X_{n+1} = b \mid X_n = c) = 1 - q$, and goes into the open air with probability q: $\mathbf{P}(X_{n+1} = a \mid X_n = c) = q$. Once she had has had sufficient fresh air she returns to the library: $\mathbf{P}(X_{n+1} = b \mid X_n = a) = 1$.

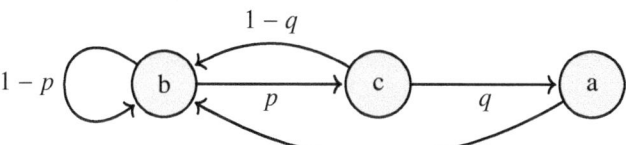

b) The transition matrix of this Markov chain is given by

$$P = \begin{matrix} & \begin{matrix} a & b & c \end{matrix} \\ \begin{matrix} a \\ b \\ c \end{matrix} & \begin{pmatrix} 0 & 1 & 0 \\ 0 & 1-p & p \\ q & 1-q & 0 \end{pmatrix} \end{matrix}.$$

We observe that the sums of the coefficients of each row is equal to 1, which means that this is a stochastic matrix.

c) A stationary law for this Markov chain is given by a row-matrix $M = (a\ b\ c)$ with positive coefficients, and sum 1, such that $M \cdot P = M$. Let us write down this system of linear equations:

$$qc = a, \quad a + (1-p)b + (1-q)c = b, \quad pb = c.$$

Then $a = qc = pqb$ and $c = pb$, and the last relation

$$pqb + (1-p)b + p(1-q)b = b$$

is automatically satisfied. The relation $a + b + c = 1$ can be written $(pq + 1 + p)b = 1$, hence $b = 1/(1 + p + pq)$. This proves that there is a unique stationary law, and that it is given by

$$\left(\frac{pq}{1+p+pq}, \frac{1}{1+p+pq}, \frac{p}{1+p+pq} \right).$$

d) By proposition 1.5.4, the entropy rate of a homogeneous stationary Markov chain is the average, for the stationary law, of the entropy of the rows of its transition matrix. These entropies are respectively equal to

$$H_a = 0,$$
$$H_b = -(1-p)\log(1-p) - p\log(p) = h(p),$$
$$H_c = -q\log(q) - (1-q)\log(1-q) = h(q),$$

where we denoted by $h(p)$ the entropy of a Bernoulli random variable with parameter p. Then

$$H(X) = aH_a + bH_b + cH_c = \frac{h(p) + ph(q)}{1 + p + pq}.$$

e) By theorem 1.5.10, this computation still holds if the Markov chain is primitive. We apply remark 1.5.11 to check that this is the case here. The cycle b → c → a → b implies that any state can be reached from any other; consequently, this Markov chain is irreducible. On the other hand, there is a length-1 cycle b → b, so that the gcd of the lengths of the cycles in each state are equal to 1; this implies that the Markov chain is aperiodic. It is therefore primitive.

Exercise 1.18. [p. 54]   a) Let us consider the quiver whose neighbors are the 64 squares of the chessboard and where arrows link each square to each of its neighbors, which are the squares a king may reach in one move. This quiver is connected; it is also aperiodic because one can go from any square to itself in two steps (make a move, and go back), or in three steps (one diagonal move, followed by one vertical move and one horizontal move). This implies that (X_n) is a primitive Markov chain.

b) Since this Markov chain is primitive, it possesses a unique stationary law. In principle, we have to look for a vector with 64 components, but the hint suggests that only three components will have distinct entries, depending only on the position of the square, according to whether it is interior, on the boundary, or on the corner.

Let us write the equations for these three real numbers c, b, i, the respective probabilities of being on the corner, on the boundary, or on the interior, of the chessboard.

4.1. Entropy and mutual information.

First of all, the sum of all probabilities is 1; since there are 4 corners, 24 boundary squares, and 36 interior squares, this gives

$$4c + 24b + 36i = 1. \tag{4.1.0.1}$$

A corner square has two neighboring squares which are boundary squares, and one neighboring square which is interior. Since boundary squares have 5 neighbors and interior squares have 8 neighbors, the stationarity relation on a corner square writes as

$$c = 2 \times \frac{1}{5}b + \frac{1}{8}i. \tag{4.1.0.2}$$

For boundary squares, there are two possibilities, since some boundary squares neighbor a corner square, and others don't. This gives two relations

$$b = \frac{1}{3}c + 2 \times \frac{1}{5}b + 2 \times \frac{1}{8}i \quad \text{and} \quad b = 2 \times \frac{1}{5}b + 3 \times \frac{1}{8}i. \tag{4.1.0.3}$$

There are three possibilities regarding interior squares, those who do not neighbor the boundary, those who neighbor the boundary but not a corner, and those who neighbor a corner. This gives three relations:

$$i = 8 \times \frac{1}{8}i, \quad i = 3 \times \frac{1}{5}b + 5 \times \frac{1}{8}i, \quad \text{and} \quad i = \frac{1}{3}c + 4 \times \frac{1}{5}b + 3 \times \frac{1}{8}i. \tag{4.1.0.4}$$

Three unknowns, seven equations, it may seem a miracle that this system can be solved at all. By elimination, we obtain immediately

$$\frac{1}{8}i = \frac{1}{5}b = \frac{1}{3}c. \tag{4.1.0.5}$$

And substituting this common value t into equation (4.1.0.1), we obtain

$$4 \times 3t + 24 \times 5t + 36 \times 8t = 1,$$

hence $t = 1/420$ and

$$c = \frac{3}{420} = \frac{1}{140}, \quad b = \frac{5}{420} = \frac{1}{84}, \quad i = \frac{8}{420} = \frac{2}{105}. \tag{4.1.0.6}$$

c) The entropy rate of the corresponding random process is given by proposition 1.5.4: $H(X) = \sum_a \mu_a H(p_{a,b})$, where a ranges over all squares, μ_a is the probability that the king is on square a for the stationary law, and $p_{a,b} = \mathbf{P}(X_{n+1} = b \mid X_n = a)$ is the transition probability from square a to square b. There are three possibilities, according to whether the square a is a corner, on the boundary, or interior.

- If a is a corner, then $\mu_a = 1/140$ and up to reordering, the vector $(p_{a,b})_b$ is equal to $(1/3, 1/3, 1/3, 0, \dots)$, so that $H((p_{a,b})_b) = \log(3)$;
- If a is on the boundary, then $\mu_a = 1/84$ and the vector $(p_{a,b})_b$ is given by $(1/5, 1/5, 1/5, 1/5, 1/5, 0, \dots)$, up to reordering, so that $H((p_{a,b})_b) = \log(5)$;

- Finally, if a is an interior square, then $\mu_a = 2/105$ and the vector $(p_{a,b})_b$ is given by $(1/8, \ldots, 1/8, 0, \ldots)$, still up to reordering, so that $H\big((p_{a,b})_b\big) = \log(8)$.

We get

$$H(X) = 4 \times \frac{1}{140} \log(3) + 24 \times \frac{1}{84} \log(5) + 36 \times \frac{2}{105} \log(8) \approx 1.917\ldots \quad (4.1.0.7)$$

Remark. — With software such as SageMath, it is possible to solve the exercise in a mechanical way. We first start by building the matrix `MKing` representing the transition probabilities:

```
# Three functions to handle the coordinates
def n2xy(n):
    return (n%8, n//8)
def xy2n(x,y):
    return x+y*8
def valid(x,y):
    return (x >= 0 and y >= 0 and x < 8 and y <8)
# Legal moves of a king
King=[]
for e in [-1,0,1]:
    for f in [-1,0,1]:
        King = King +[(e,f)]
King.remove((0,0))
# The transition matrix for the king's moves
MKing = []
for n in range(64):
    MKing.append([0]*64)
    x,y=n2xy(n)
    Kxy = [].copy()
    for dx, dy in King:
        if valid(x+dx,y+dy):
            Kxy.append (xy2n(x+dx,y+dy))
    for d in Kxy:
        MKing[-1][d] = 1/len(Kxy)
MKing=matrix(QQ,MKing)
```

The command

```
evKing = MKing.eigenvalues()
```

computes the eigenvalues of the matrix `MKing`; the first four are

$$\left[1, -\frac{1}{5}, -0.4910211476797844, -0.3316809492706161\right].$$

In particular, 1 is the greatest eigenvalue of this primitive stochastic matrix. We then compute the corresponding normalized left-eigenvector:

4.2. Coding. 171

```
# Left eigenvector for the eigenvalue 1
EVKing = MKing.eigenvectors_left()[0][1][0]
EVKing = EVKing/sum(EVKing)
```

Its first coordinates are:

$$\frac{1}{140}, \frac{1}{84}, \frac{1}{84}, \frac{1}{84}, \frac{1}{84}, \frac{1}{84}, \frac{1}{84}, \frac{1}{140}, \frac{1}{84}, \frac{2}{105}, \cdots$$

Finally, we compute the entropy rate (the function H computes Shannon's entropy of a row vector):

```
H = lambda p : sum(map(lambda x: 0
    if x <= 0 else -x*log(x),p))
HKing = RR(sum([EVKing[i]*H(MKing[i])
    for i in range(len(EVKing))]))
```

and we again find the value 1.91713109752358 that we had computed.

4.2. Coding

Exercise 2.1. [p. 89]   *a*) Since the code word 0 is a prefix of the code word 01, this code is not a prefix code.

b) Let $a = a_1 \ldots a_m$ and $b = b_1 \ldots b_n$ be words that have the same code; let us prove, by induction on their length, that they are equal. If one of them is empty, so is the other, so we may assume that they are not empty; we then write $a = a_1 a'$ and $b = b_1 b'$, so that the code of a is $C(a_1)C(a')$ and that of b is $C(b_1)C(b')$. If $a_1 = b_1$, then a' and b' have the same code, so are equal by induction; this implies $a = ab$. Let us now assume that $a_1 \neq b_1$ and, to fix ideas, that the code of a_1 is 0 while that of b_1 is 01. In this case, we have $0C(a') = 01C(b')$, hence $C(a') = 1C(b')$; in particular, a' is not empty ($m \geqslant 2$) and the code of a_2 starts with a 1, which is absurd since it is either 0 or 01.

A simpler proof would read the words from right to left, which amounts to pretending that the two code words of our code are 0 and 10. This new code is a prefix code, hence is uniquely decodable, so that our code is uniquely decodable.

Exercise 2.2. [p. 89]   *a*) By definition of independence, we have

$$P(Y_n = (a_1, \ldots, a_n)) = P(X_1 = a_1) \cdots P(X_n = a_n)$$

for every $(a_1, \ldots, a_n) \in A^n$. Since the process (X_n) is stationary, all the variables X_n follow the same law, that of X_1, and we have

$$P(Y_n = (a_1, \ldots, a_n)) = P(X_1 = a_1) \cdots P(X_1 = a_n).$$

The entropy of Y_n is $nH(X_1)$.

b) By Shannon's theorem, we have

$$H_D(Y_n) \leq \mathbf{E}\big(\ell(C_n(Y_n))\big) < H_D(Y_n) + 1.$$

Consequently,

$$\frac{1}{n}H_D(Y_n) \leq L_n < \frac{1}{n}H_D(Y_n) + \frac{1}{n}.$$

When $n \to +\infty$, the left- and right-hand sides of this inequality converge to the entropy rate $H(X)$ of the random process (X_n). This implies that (L_n) converges to that entropy rate.

Exercise 2.3. [p. 90]   a) We just need to apply the definition of entropy, $H(X) = -\sum p(a) \log\big(p(a)\big)$. For example, using SageMath:

```
h2 = lambda p:-p*log(p,2)
H2 = lambda p: sum (map(h2,p))
p = [0.49, 0.26, 0.12, 0.04, 0.04, 0.03, 0.02]
```

we obtain that the base 2 entropy of this random variable is 2.0127...

b) We complete the table by indicating the length of the code word of each symbol, as given in Shannon's method.

a	b	c	d	e	f	g
0.49	0.26	0.12	0.04	0.04	0.03	0.02
2	2	4	5	5	6	6

We then successively add the fractions $2^{-\ell}$ and isolate the relevant part of the binary expansion.

a 2	0.	00
b 2	0.01	01
c 4	0.10	1000
d 5	0.1001	10010
e 5	0.10011	10011
f 6	0.10100	101000
g 6	0.101001	101001

Using the SageMath code,

```
lS = lambda p:ceil(-ln(p)/ln(2.0))
pS = list(map(lS,p))
lpS = sum (pS[i]*p[i] for i in range(len(p)))
```

we obtain that the average length of this code is $0.49 \times 2 + 0.226 \times 2 + \cdots = 2.68$.

4.2. Coding.

c) Huffman's method starts from the initial probabilities

a	b	c	d	e	f	g
0.49	0.26	0.12	0.04	0.04	0.03	0.02

and combines two symbols with minimal probabilities, here f and g; the sum of their probabilities is 0.04, and we get a new alphabet:

a	b	c	d	e	fg
0.49	0.26	0.12	0.04	0.04	0.05

Going on, we now combine d and e:

a	b	c	de	fg
0.49	0.26	0.12	0.08	0.05

then the groups de and fg:

a	b	c	defg
0.49	0.26	0.12	0.13

the symbol c and the group defg:

a	b	cdefg
0.49	0.26	0.25

and finally the symbol b and the group cdefg:

a	bcdefg
0.49	0.51

We then decide to code a as 0, bcdefg as 1; b as 10 and cdefg as 11; c as 110 and defg as 111; de as 1110 and fg as 1111; d as 11100 and e as 11101, f as 11110 and *g* as 11111, hence the following code:

a	b	c	d	e	f	g
0.49	0.26	0.12	0.04	0.04	0.03	0.02
0	10	110	11100	11101	11110	11111
1	2	3	5	5	5	5

The SageMath instructions

```
pH = [1,2,3,5,5,5,5,5]
lpH = sum (pH[i]*p[i] for i in range(len(p)))
```

compute the average length of this code, and we obtain 2.02.

Exercise 2.4. [p. 90]   a) If $q(a) = 1/2^n$, then $\log_2(q(a)) = -n$ and $\ell(C(a)) = n$. Consequently,

$$E_q\big(\ell(C(a))\big) = \sum_a \big(-q(a)\log_2(q(a))\big) = H_2(q).$$

Given Shannon's inequality, the code C is optimal with respect to the law q. The average length of a code word is

$$E\big(\ell(C(X))\big) = \sum_a p(a)\ell(C(a)) = -\sum_a p(a)\log_2(q(a)).$$

We guess a relation with the divergence $D(p,q)$, so let us introduce a term $\log_2(p(a))$:

$$E\big(\ell(C(X))\big) = -\sum_a p(a)\log_2\left(\frac{q(a)}{p(a)}\right) - \sum_a p(a)\log_2(p(a)) = D(p,q) + H(p).$$

This shows in this case how the divergence $D(p,q)$ of q with respect to p measures how the average length of a code word exceeds Shannon's bound.

b) We do the same computation, using the bounds

$$-\log_2(q(a)) \leq \ell(C(a)) < 1 - \log_2(q(a)).$$

Then

$$E\big(\ell(C(X))\big) = \sum_a p(a)\ell(C(a)) \geq -\sum_a p(a)\log_2 q(a) = D(p,q) + H(p).$$

Similarly,

$$E\big(\ell(C(X))\big) = \sum_a p(a)\ell(C(a))$$
$$\leq \sum_a p(a) + D(p,q) + H(p)$$
$$= 1 + D(p,q) + H(p).$$

Since $\sum p(a) = 1$, there is at least one symbol for which $p(a) > 0$; then $\ell(C(a)) < 1 - \log_2(q(a))$, and the final upper bound is strict.

4.2. Coding.

Exercise 2.5. [p. 90]   *a*) Code I is prefix, but not code II.
These two codes are uniquely decodable. For code I, this follows from it being a prefix code. For code II, we may observe that it is obtained from code I by writing the symbols from right to left; to decode a text, it is enough to read it in reverse, to decode it as if it were coded by code I, and to write the result in reverse.

b) We know that

$$I(U, V) = H(U) + H(V) - H(U, V) = H(V) - H(V \mid U).$$

We have $\mathbf{P}(U = \text{true}) = 0.4$ and $\mathbf{P}(U = \text{false}) = 0.6$. Consequently,

$$H(U) = -0.4 \log(0.4) - 0.6 \log(0.6) \approx 0.971.$$

Let us start with code I. In this case, $V = U$ since a code words starts with 1 precisely when the symbol is *a*. Consequently, $I(U, V) = H(U) \approx 0.971$.

Let us now answer this question for code II. We have $\mathbf{P}(V = \text{true}) = 1$; the random variable V is certain, hence is independent of U, hence $I(U, V) = 0$.

c) The code words for codes I and II have the same length, so that the results will be the same in both cases.
We have

$$\mathbf{E}\big(\ell(C(X))\big) = \mathbf{P}(X = a)\ell(C(a)) + \cdots + \mathbf{P}(X = d)\ell(C(d))$$
$$= 0.4 \times 1 + 0.3 \times 2 + 0.2 \times 3 + 0.1 \times 4 = 2.$$

According to Shannon's theorem, we have $\mathbf{E}\big(\ell(C(X))\big) \geqslant H_2(X)$, the base 2 entropy of X, and there exists a code of average length at most $1 + H_2(X)$. On the other hand, the entropy of X is given by

$$H_2(X) = -0.4 \log_2(0.4) - \cdots - 0.1 \log_2(0.1) \approx 1.846.$$

Thus the average length lies within the bounds of Shannon's theorem (as it should).

d) Let us construct an optimal code for the variable U using Huffman's method. We start from the set $A = \{a, b, c, d\}$ and the probabilities given by the table

a	b	c	d
0.4	0.3	0.2	0.1

We first combine the two symbols *c* and *d* to a symbol *cd* with probability 0.3. This gives the table:

a	b	cd
0.4	0.3	0.3

We then combine the two symbols b and cd into a symbol cd with probability 0.6, hence the table

a	bcd
0.4	0.6

Then we code a as 0, and bcd as 1; then b as 10 and cd as 11; then c as 110 and d as 111. We deduce the following code:

X	p_X	code III
a	0.4	0
b	0.3	10
c	0.2	110
d	0.1	111

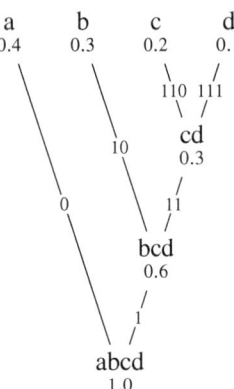

Its average length is given by

$$\mathbf{E}\big(\ell(C_{III}(X))i\big) = 0.4 \times 1 + 0.3 \times 2 + 0.2 \times 3 + 0.1 \times 3 = 1.9.$$

This is very close to the entropy that we had computed in the preceding question.

Exercise 2.6. [p. 91]   *a*) Since the exponential function is strictly increasing and $t > 0$, we have

$$\mathbf{P}(X \geq a) = \mathbf{P}(tX \geq ta) = \mathbf{P}(e^{tX} \geq e^{ta}) \leq \mathbf{E}(e^{tX})e^{-ta},$$

by Markov's inequality for the random variable e^{tX}.

Remark. — Chernoff's inequality applies in more restricted situations than Markov's inequality (which only needs the existence of expectation) or than the Bienaymé–Chebyshev inequality (which requires the variance to exist), but it furnishes finer information.

b) For $k \in \{0, \ldots, n\}$, we have $\mathbf{P}(S = k/n) = \binom{n}{k} p^k (1-p)^{n-k}$, the probability that among $X_1, \ldots, X_n$, we get k times the value 1. One has $\mathbf{P}(S = x) = 0$ if x is not of this form. Therefore,

$$\mathbf{E}(e^{tS}) = \sum_{k=0}^{n} \binom{n}{k} p^k (1-p)^{n-k} e^{tk/n} = \big((1-p) + p e^{t/n}\big)^n,$$

according to the binomial identity.

c) We have

$$D(q, p) = q \log\left(\frac{q}{p}\right) + (1-q) \log\left(\frac{1-q}{1-p}\right)$$
$$= q \log(q) + (1-q) \log(1-q) - q \log(p) - (1-q) \log(1-p).$$

4.2. Coding.

d) Let us write $\mathbf{E}(e^{tS})e^{-tq} = f(t/n)^n$, with

$$f(x) = \big((1-p) + pe^x\big)e^{-qx} = (1-p)e^{-qx} + pe^{(1-q)x}.$$

For $q > 1$, the function f is strictly decreasing on $\mathbf{R}$, and goes to 0 at $+\infty$; we then have $\inf_t \mathbf{E}(e^{tS})e^{-tq} = 0$. Applying Chernoff's inequality, we get $\mathbf{P}(S \geqslant q) = 0$, which is obvious since S takes its values in the interval $[0; 1]$. For $q = 1$, we obtain the boring inequality $\mathbf{P}(S \geqslant 1) \leqslant p$. If $q \leqslant 0$, the function f is strictly increasing on $\mathbf{R}$, hence $\inf_{x>0} f(x) = 1$, and we get $\mathbf{P}(S \geqslant q) \leqslant 1$, an equally boring inequality.

Let us now assume that $0 < q < 1$. The function f is infinitely differentiable, and its derivative is given by

$$f'(x) = -(1-p)qe^{-qx} + p(1-q)e^{(1-q)x}.$$

It follows that f is strictly decreasing on $]-\infty; \xi]$ and strictly increasing on $[\xi; +\infty[$, where ξ is the real number defined by

$$e^\xi = \frac{1-p}{p} \times \frac{q}{1-q}.$$

Consequently,

$$\begin{aligned} f(\xi) &= \big((1-p) + pe^\xi\big)e^{-q\xi} \\ &= (1-p)\Big(1 + \frac{q}{1-q}\Big)p^q(1-p)^{-q}q^{-q}(1-q)^q \\ &= (1-p)^{1-q}p^q/q^q(1-q)^{1-q} = \exp\big(-D(q,p)\big). \end{aligned}$$

Since $q > p$, we have $1/p > 1/q$, then $1/p - 1 > 1/q - 1$ that is, $(1-p)/p > (1-q)/q$; in other words, $\xi > 0$. It follows that we may apply the Chernoff inequality and we obtain

$$\mathbf{P}(S \geqslant q) \leqslant \mathbf{E}(e^{n\xi S})e^{-n\xi q} = \exp\big(-nD(q,p)\big).$$

Since $q > p$, we have $D(q,p) > 0$, and this furnishes an exponential decay of $\mathbf{P}(S \geqslant q)$.

e) Let us set $X'_k = 1 - X_k$; it is a Bernoulli variable with parameter $1 - p$. The sequence $(X'_1, \ldots, X'_n)$ is independent. Setting $S' = (X'_1 + \cdots + X'_n)/n$, we have $S' = 1 - S$, and the previous question implies that $\mathbf{P}(S' \geqslant 1-q) \leqslant \exp\big(-nD(1-q, 1-p)\big)$. Consequently,

$$\mathbf{P}(S \leqslant q) \leqslant \exp\big(-nD(1-q, 1-p)\big) = \exp\big(-nD(q,p)\big).$$

Exercise 2.7. [p. 91]   *a)* We start from the definition

$$D(q,p) = -q\log(p) - (1-q)\log(1-p) + q\log(q) + (1-q)\log(1-q)$$

and we differentiate twice with respect to p:

$$\partial_p D(q,p) = -q\frac{1}{p} + (1-q)\frac{1}{1-p},$$
$$\partial_p^2 D(q,p) = q\frac{1}{p^2} + (1-q)\frac{1}{(1-p)^2}.$$

In particular, $D(q,p)$ is a convex function of the variable p, and it vanishes for $p=q$. Since the function $x \mapsto 1/x^2$ is convex, we have

$$\partial_p^2 D(q,p) \geq \frac{1}{\big(qp + (1-q)(1-p)\big)^2} \geq 1,$$

given that $qp + (1-q)(1-p) \leq \sup(p, 1-p) \leq 1$. By the Taylor formula of order 2, it follows that

$$D(q,p) \geq \frac{1}{2}(q-p)^2.$$

b) Using exercise 2.6, we then have

$$\mathbf{P}(S - pn \geq \varepsilon pn) \leq \exp\big(-nD((1+\varepsilon)p, p)\big)$$
$$\leq \exp\big(-n\frac{1}{2}(\varepsilon p - p)^2\big) = \exp\big(-\frac{1}{2}n\varepsilon^2 p^2\big),$$

and a similar inequality for $\mathbf{P}(S - pn \leq -\varepsilon pn)$:

$$\mathbf{P}(S - pn \leq -\varepsilon pn) \leq \exp\big(-nD((1-\varepsilon)p, p)\big) \leq \exp\big(-\frac{1}{2}n\varepsilon^2 p^2\big).$$

The required inequality follows from this.

Exercise 2.8. [p. 91]   a) This channel is symmetric. For every $a \in A$, we have

$$H(Y \mid X = a) = -2p\log(p) - (1-2p)\log(1-2p).$$

It then follows from the formula of example 2.5.4, 3, that we have

$$I(C) = \log(d) + 2p\log(p) + (1-2p)\log(1-2p).$$

b) Let X and Y be random variables with values in A such that $X \sim_C Y$; let $u = (u_a)$ be the law of X. We have $I(X, Y) = H(Y) - H(X \mid Y)$. The expression $H(X \mid Y)$ is equal to $-2p\log(p) - (1-2p)\log(1-2p)$, while $H(Y)$ is at most $\log(d)$, and equality holds if and only if the law of Y is uniform, that is, $\mathbf{P}(Y = b) = 1/d$ for all $b \in A$. We have

$$\mathbf{P}(Y = b) = (1 - 2p)u_b + pu_{b-1} + pu_{b+1}.$$

If the law of X is itself uniform, we have $u_b = 1/d$ for all b, and this implies $\mathbf{P}(Y = b) = 1/d$ for all b, so that the law of Y is uniform. (In some cases, there may be non-uniform laws for X that make the law of Y uniform; this happens for example for $d = 8$ and $p = 1/4$.)

Exercise 2.9. [p. 91] Let X and Y be random variables adapted to this channel: $\mathbf{P}(Y = 0 \mid X = 0) = 1$ and $\mathbf{P}(Y = 0 \mid X = 1) = p$. Let $u = \mathbf{P}(X = 1)$; we then have

$$\mathbf{P}(Y = 1) = \mathbf{P}(X = 0)\mathbf{P}(Y = 1 \mid X = 0) + \mathbf{P}(X = 1)\mathbf{P}(Y = 1 \mid X = 1) = u(1-p),$$

and $\mathbf{P}(Y = 0) = 1 - u(1-p)$. Then

$$H(Y) = -\big(1 - u(1-p)\big)\log\big(1 - u(1-p)\big) - u(1-p)\log\big(u(1-p)\big),$$

while

$$\begin{aligned}
H(Y \mid X) &= \mathbf{P}(X=0)H(Y \mid X=0) + \mathbf{P}(X=1)H(Y \mid X=1) \\
&= (1-u) \times 0 + u \times \big(-p\log(p) - (1-p)\log(1-p)\big) \\
&= -up\log(p) - u(1-p)\log(1-p).
\end{aligned}$$

Consequently,

$$\begin{aligned}
I(X,Y) &= H(Y) - H(Y \mid X) \\
&= -\big(1 - u(1-p)\big)\log\big(1 - u(1-p)\big) - u(1-p)\log\big(u(1-p)\big) \\
&\quad + up\log(p) + u(1-p)\log(1-p) \\
&= -\big(1 - u(1-p)\big)\log\big(1 - u(1-p)\big) - u(1-p)\log(u) + up\log(p).
\end{aligned}$$

Writing $F(u, p)$ for this expression, let us study its behavior as a function of u; we omit p from the notation. To simplify we assume for the moment that $0 < p < 1$. We compute

$$\begin{aligned}
F'(u) &= (1-p)\log\big(1 - u(1-p)\big) + p\log(p) - (1-p)\log(u) \\
&= (1-p)\log\big(\frac{1}{u} - (1-p)\big) + p\log(p),
\end{aligned}$$

which proves that the function F' is strictly decreasing on $[0;1]$. We also have $F'(0^+) = +\infty$ and $F'(1) = \log(p) < 0$. Consequently, there exists a unique real number $u_p \in \,]0;1[$ such that $F'(u_p) = 0$; the function F is strictly increasing on $[0; u_p]$ and strictly decreasing on $[u_p; 0]$, and we have $I(C) = F(u_p, p)$. Actually, u_p can be computed explicitly: we get

$$\frac{1}{u_p} = (1-p) + \exp\big(-\frac{p}{1-p}\log(p)\big).$$

We check that the transmission capacity of this channel decreases with p. It tends to $\log(2)$ when $p \to 0$ (the channel "converges" to a channel without noise!) and to 0 when $p \to 1$ (the channel only passes 0).

Exercise 2.10. [p. 92]   *a)* We assume that the channels C' and C'' are independent. Write A for the input alphabet of the channel C', A' for its output alphabet, which is also the input alphabet of the channel C'', and B for the output alphabet

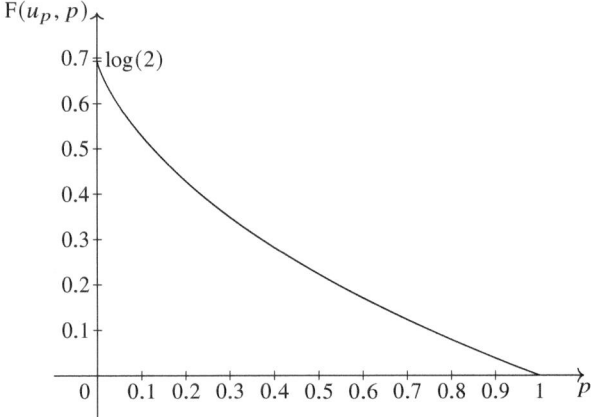

Fig. 4.1 Graph of the function $p \mapsto F(u_p, p)$

of C″. Then, for every $a \in A$ and $b \in B$, we have

$$p_C(a \mid b) = \sum_{a'} p_{C'}(a \mid a') p_{C''}(a' \mid b).$$

The matrix of transmission probabilities of the channel C is the product of the matrices of transmission probabilities of the channels C′ and C″.

b) Let X, Z be discrete random variables with values in A, B such that $X \sim_C Z$. Let us introduce a random variable Y, inbetween X and Z, revealing the result of the transmission of X in the channel C′. We thus have $X \sim_{C'} Y$ and $Y \sim_{C''} Z$. By definition, $I(X, Y) \leq I(C')$ and $I(Y, Z) \leq I(C'')$.

The goal is to bound $I(X, Z)$ from above. We remark that X and Z are conditionally independent with respect to Y: the memoryless channel C″ receives Y and passes Z; knowing X, the source of the channel C′ gives no additional information on Z. By the data processing inequality (theorem 1.3.11), we then have $I(X, Z) \leq I(X, Y) = I(C')$. As we saw in the proof of this theorem, this inequality results from the relations

$$I(X, (Y, Z)) = I(X, Z \mid Y) + I(X, Y) = I(X, Y \mid Z) + I(X, Z),$$

the vanishing of $I(X, Z \mid Y)$ (since $X \perp_Y Z$) and the positivity of $I(X, Y \mid Z)$. By symmetry, we also have

$$I((X, Y), Z) = I(X, Z \mid Y) + I(Y, Z) = I(Y, Z \mid X) + I(X, Z),$$

which implies
$$I(X, Z) = I(Y, Z) - I(Y, Z \mid X) \leq I(Y, Z).$$

In particular, $I(C) \leq \inf(I(C'), I(C''))$: the channel C cannot do more than what any of the channels C′ and C″ do.

4.2. Coding.

c) Let A be the matrix of transmission probabilities of the code C:

$$A = \begin{pmatrix} 1-p & p \\ p & 1-p \end{pmatrix}.$$

The matrix of transmission probabilities of the code C_n is the n-th power A^n of A. We wish to prove that A^n is the matrix of transmission probabilities of a symmetric binary transmission channel with parameter: $p_n = (1 - (1-2p)^n)/2$, that is,

$$A^n = \begin{pmatrix} 1-p_n & p_n \\ p_n & 1-p_n \end{pmatrix}.$$

We can prove this by induction, or observe that A^n is symmetric (being a power of a symmetric matrix) and that $PA^nP = A^n$, where $P = \begin{pmatrix} 0 & 1 \\ 1 & 0 \end{pmatrix}$. In particular, it is the matrix of transmission probabilities of a symmetric binary transmission channel of *some* parameter p_n. To compute the value of the real number p_n, we observe that

$$\det(A^n) = \det(A)^n = \big((1-p)^2 - p^2\big)^n = (1-2p)^n = 1 - 2p_n.$$

Consequently, $p_n = (1 - (1-2p)^n)/2$, as claimed.

Let us assume that $0 < p < 1$, so that $|1 - 2p| < 1$. When n tends to infinity, we have $(1-2p)^n \to 0$ hence $p_n \to 1/2$. The transmission capacity of the channel C_n is equal to $1 - h_2(p_n)$, so it converges to $1 - h_2(1/2) = 0$ when $n \to \infty$.

Exercise 2.11. [p. 92]   a) Assume that the two random variables $X = (X', X'')$ and $Y = (Y', Y'')$ are related by the channel C, that is, that we have

$$\mathbf{P}(Y = (b', b'') \mid X = (a', a'')) = p(b' \mid a') \, p(b'' \mid a'').$$

Then

$$\mathbf{P}(Y' = b' \mid X' = a') = \sum_{b'' \in B''} \mathbf{P}(Y' = b', Y'' = b'' \mid X' = a')$$

$$= \sum_{b'' \in B''} \mathbf{P}(X' = a')^{-1} \mathbf{P}(Y' = b', Y'' = b'', X' = a')$$

$$= \mathbf{P}(X' = a')^{-1}$$
$$\times \sum_{b'' \in B''} \sum_{a'' \in A''} \mathbf{P}(Y' = b', Y'' = b'', X' = a', X'' = a'')$$

$$= \mathbf{P}(X' = a')^{-1} \sum_{a'' \in A''} \mathbf{P}(X' = a', X'' = a'')$$
$$\times \sum_{b'' \in B''} \mathbf{P}(Y' = b', Y'' = b'' \mid X' = a', X'' = a'')$$

$$= \mathbf{P}(X' = a')^{-1} \sum_{a'' \in A''} \mathbf{P}(X' = a', X'' = a'')$$
$$\times \sum_{b'' \in B''} p(b' \mid a') p(b'' \mid a'')$$

$$= \mathbf{P}(X' = a')^{-1} \sum_{a'' \in A''} \mathbf{P}(X' = a', X'' = a'') p(b' \mid a')$$

$$= p(b' \mid a').$$

This proves that the random variables X' and Y' are related by the channel C', and $I(X', Y') \leq I(C')$. Similarly, the random variables X'' and Y'' are related by the channel C'', and $I(X'', Y'') \leq I(C'')$.

b) Intuitively, given Y', the knowledge of X'' does not give any information about X'. Actually, for any $a' \in A'$, $b' \in B'$ and $a'' \in A''$, we have

$$\begin{aligned}
\mathbf{P}(Y' &= b', X'' = a'' \mid X' = a') \\
&= \mathbf{P}(X'' = a'' \mid X' = a') \mathbf{P}(Y' = b' \mid X' = a', X'' = a'') \\
&= \mathbf{P}(X'' = a'' \mid X' = a') \times \sum_{b'' \in B''} \mathbf{P}(Y' = b', Y'' = b'' \mid X' = a', X'' = a'') \\
&= \mathbf{P}(X'' = a'' \mid X' = a') \sum_{b'' \in B''} p(b' \mid a') p(b'' \mid a'') \\
&= \mathbf{P}(X'' = a'' \mid X' = a') p(b' \mid a') \\
&= \mathbf{P}(X'' = a'') \mathbf{P}(Y' = b' \mid X' = a'),
\end{aligned}$$

which proves that Y' and X'' are conditionally independent given X'. By symmetry, the proof of the other conditional independence is identical.

c) Let us observe that

$$I(X, Y) = I\big((X', X''), (Y', Y'')\big) = H(Y', Y'') - H(Y', Y'' \mid X', X'').$$

The relation

$$\mathbf{P}(Y' = b', Y'' = b'' \mid X' = a', X'' = a'') = p(b' \mid a') \, p(b'' \mid a'')$$

implies that the random variables Y' and Y'' are conditionally independent given X', X'', so that

$$H(Y', Y'' \mid X', X'') = H(Y' \mid X', X'') + H(Y'' \mid X', X'').$$

Since Y' and X'' are conditionally independent given X', we also have

$$H(Y' \mid X', X'') = H(Y', X'' \mid X') - H(X'' \mid X') = H(Y' \mid X').$$

Similarly, we have
$$H(Y'' \mid X', X'') = H(Y'' \mid X'').$$

Consequently,

4.2. Coding.

$$I(X, Y) = H(Y', Y'') - H(Y' \mid X') - H(Y'' \mid X'')$$
$$= H(Y') + H(Y'') - I(Y', Y'') - H(Y' \mid X') - H(Y'' \mid X'')$$
$$= I(X', Y') + I(X'', Y'') - I(Y', Y'').$$

The first term is at most $I(C')$ and the second is at most $I(C'')$; the third term is positive, and it vanishes if and only if Y' and Y'' are independent. This proves that $I(X, Y) \leqslant I(C') + I(C'')$, with equality if and only if Y' and Y'' are independent. In particular, $I(C) \leqslant I(C') + I(C'')$.

Let us now assume that X' and X'' are independent, and that their laws are such that $I(X', Y') = I(C')$ and $I(X'', Y'') = I(C'')$. Then Y' and Y'' are independent, since

$$\mathbf{P}(Y' = b', Y'' = b'') = \sum_{a' \in A'} \sum_{a'' \in A''} \mathbf{P}(X' = a', X'' = a'')$$
$$\times \mathbf{P}(Y' = b', Y'' = b'' \mid X' = a', X'' = a'')$$
$$= \sum_{a' \in A'} \sum_{a'' \in A''} \mathbf{P}(X' = a') \mathbf{P}(X'' = a'') p(b' \mid a') p(b'' \mid a'')$$
$$= \sum_{a' \in A'} \mathbf{P}(X' = a') p(b' \mid a') \sum_{a'' \in A''} \mathbf{P}(X'' = a'') p(b'' \mid a'')$$
$$= \mathbf{P}(Y' = b') \mathbf{P}(Y'' = b'').$$

We then have $I(X, Y) = I(X', Y') + I(X'', Y'') = I(C') + I(C'')$. This proves that the transmission capacity of the channel C is equal to $I(C') + I(C'')$, and that this capacity is reached by the probability law on $A' \times A''$ which is the product of laws on A' and A'' that reach the transmission capacity of C' and C''.

Exercise 2.12. [p. 93]   *a)* Let $a \in A''$ and $b \in B'$. The sum of all conditional probabilities $p_C(x \mid a)$ is equal to 1; we thus have

$$p_C(b \mid a) = 1 - \sum_{x \in B' \cup B'' - \{b\}} p_C(x \mid a)$$
$$= 1 - \sum_{x \in B' - \{b\}} p_C(b \mid a) - \sum_{x \in B''} p_C(x \mid a).$$

All terms $p_C(x \mid a)$ are positive; moreover, we have $\sum_{x \in B''} p_C(x \mid a) = 1$. We thus obtain $p_C(b \mid a) \leqslant 0$, hence $p_C(b \mid a) = 0$.

An analogous argument proves that $p_C(b \mid a) = 0$ if $b \in B''$ and $a \in A'$.

b) Let X, Y be random variables with values in A and B respectively, adapted to the channel C: $\mathbf{P}(Y = b \mid X = a) = p_C(b \mid a)$. Let U be the random variable equal to 1 if $X \in A'$ and to 0 if $X \in A''$. When $U = 1$, we also have $Y \in B'$; when $U = 0$, we have $Y \in B''$. Set $p = \mathbf{P}(U = 1)$. We compute $I(X, Y)$ by conditioning. Since U is certain given X or Y, we have

$$I(X, Y) = H(X) + H(Y) - H(X, Y)$$
$$= H(X, U) + H(Y, U) - H(X, Y, U)$$
$$= H(U) + H(X \mid U) + H(Y \mid U) - H(X, Y \mid U)$$
$$= H(U) + I(X, Y \mid U).$$

On the other hand,

$$I(X, Y \mid U) = \mathbf{P}(U = 0)I(X, Y \mid U = 0) + \mathbf{P}(U = 1)I(X, Y \mid U = 1).$$

Conditioned to the even $U = 0$, the pair (X, Y) behaves as a pair of random variables adapted to the channel C', so that $I(X, Y \mid U = 0) \leq I(C'')$. Similarly, $I(X, Y \mid U = 1) \leq I(C')$. Consequently,

$$I(X, Y) \leq h(p) + (1 - p)I(C'') + pI(C')$$
$$= (1 - p)(I(C'') - \log(1 - p)) + p(I(C') - \log(p)).$$

Let us denote by $f(p)$ the right-hand side of this inequality; it is a continuous function of $p \in [0; 1]$, strictly concave (as the sum of the strictly concave function $h(p)$ and of an affine function). It is equal to $I(C'')$ when $p = 0$ and to $I(C')$ when $p = 1$, and reaches its maximum at a unique point $p \in [0; 1]$. Observe that f is differentiable on the open interval $]0; 1[$ and that its derivative is given by

$$f'(p) = I(C') - \log(p) - 1 - I(C'') + \log(1 - p) + 1$$
$$= I(C') - I(C'') - \log(p) + \log(1 - p).$$

The function $p \mapsto -\log(p) + \log(1 - p)$ is strictly decreasing, because its derivative, given by $p \mapsto -1/p - 1/(1 - p)$, is strictly negative. On the other hand, f' converges to $+\infty$ at 0, and to $-\infty$ at 1; consequently, there is a unique real number p such that $f'(p) = 0$. The function f is strictly increasing on $[0; p]$, and strictly decreasing on $[p; 1]$, and we have $\sup(f) = f(p)$.

To compute p, we deduce from $f'(p) = 0$ that $\log(1/p - 1) = I(C'') - I(C')$, so that $1/p - 1 = e^{I(C'')}/e^{I(C')}$ and

$$p = \frac{e^{I(C')}}{e^{I(C')} + e^{I(C'')}}.$$

To simplify the notation, let us set $\gamma' = e^{I(C')}$ and $\gamma'' = e^{I(C'')}$. Then we have

$$I(C') - \log(p) = \log(\gamma') - \log\left(\frac{\gamma'}{\gamma' + \gamma''}\right) = \log(\gamma' + \gamma'')$$

and, by symmetry,

$$I(C'') - \log(1 - p) = \log(\gamma' + \gamma'').$$

Therefore,

4.2. Coding.

$$f(p) = p\big(I(C') - \log(p)\big) + (1-p)\big(I(C'') - \log(1-p)\big)$$
$$= p\log(\gamma' + \gamma'') + (1-p)\log(\gamma' + \gamma'')$$
$$= \log(\gamma' + \gamma'').$$

We thus have $I(X, Y) \leq \log(\gamma' + \gamma'')$, hence

$$e^{I(C)} \leq e^{I(C')} + e^{I(C'')}.$$

To prove that equality holds, let us choose independent random variables X' and X'', with values in A' and A'' respectively, that reach the capacities $I(C')$ and $I(C'')$. Let U be a Bernoulli variable with parameter p (that is, $\mathbf{P}(U = 1) = p$ and $\mathbf{P}(U = 0) = 1 - p$), independent of X', X''. Let us define a random variable X by $X = X'$ if $U = 1$, and $X = X''$ if $U = 0$; let us also consider a random variable Y such that $X \sim_C Y$. The preceding computation shows that $I(X, Y) = f(p) = \log\left(e^{I(C')} + e^{I(C'')}\right)$, so that $I(C) = \log\left(e^{I(C')} + e^{I(C'')}\right)$, a capacity which is reached by the law of the random variable X.

Exercise 2.13. [p. 93]   *a)* We assume that the symbol 0 costs c_0 and the symbol 1 costs c_1; by assumption, we have $p(1 \mid 0) = p(0 \mid 1) = p$. Let (X, Y) be a pair of random variables such that $X \sim_C Y$. Set $u = \mathbf{P}(X = 0)$ and $v = \mathbf{P}(Y = 0)$; as we saw when computing the capacity of a symmetric binary transmission channel, we have $H(Y \mid X) = h(p)$ and

$$v = (1-p)\mathbf{P}(X=0) + p\mathbf{P}(X=1) = (1-p)u + p(1-u) = u + p - 2up,$$

so that

$$I(X, Y) = h\big(u(1-2p) + p\big) - h(p),$$

an expression to maximize. To simplify, we assume $p \leq 1/2$. Then the function $u \mapsto u(1-2p) + p$ is affine increasing, its image is $[p; 1-p]$ for $u \in [0; 1]$. Moreover, $I(X, Y)$ is maximal when $u = 1/2$, that is, when X is uniform and then Y is uniform; then we have $I(X, Y) = 1 - h(p)$, the capacity of the transmission channel C.

However, we need to take the cost condition into account. We have $\mathbf{E}\big(c(X)\big) = uc_0 + (1-u)c_1$, so that the condition $\mathbf{E}\big(c(X)\big) \leq \gamma$ can be rewritten $(c_0 - c_1)u \leq \gamma - c_1$.

If this condition holds for $u = 1/2$, that is, if $(c_0 + c_1)/2 \leq \gamma$, then $I(C, \gamma) = 1 - h(p)$ for $\gamma \geq (c_0 + c_1)/2$. This is in particular the case when c_0, c_1 are both $\leq \gamma$.

On the other hand, if $\gamma < \sup(c_0, c_1)$ then, the cost of each symbol being greater than the authorized cost, the cost condition cannot be satisfied. The upper bound defining $I(C, \gamma)$ is taken in $\mathbf{R}_+$, hence $I(C, \gamma) = 0$.

Let us now assume that $(c_0 + c_1)/2 > \gamma$ and $\inf(c_0, c_1) \leq \gamma$. Set $u_0 = (c_1 - \gamma)/(c_1 - c_0)$. There are two cases, according to $c_0 < c_1$ or $c_1 < c_0$:

- If $c_0 < c_1$, then the cost condition rewrites as $u \leq u_0$; in this case, we have $u_0 < 1/2$. Consequently, u_0 is the value of u which maximizes $I(X, Y)$ while keeping the cost condition satisfied, and we get

$$I(C, \gamma) = h\left(\frac{c_1 - \gamma}{c_1 - c_0}(1 - p) + \frac{\gamma - c_0}{c_1 - c_0}p\right) - h(p).$$

- If $c_0 > c_1$, then the cost condition becomes $u \geq u_0$. Since $u_0 > 1/2$ in this case, it is the value that makes $I(X, Y)$ maximal for the given cost condition, and we obtain the same formula for $I(C, \gamma)$.

b) The computations are similar. By symmetry, we have

$$H(Y \mid X) = h(q, p, \ldots, p) = -q \log(q) - (d - 1)p \log(p).$$

Set $u_j = \mathbf{P}(X = j)$ and $v_j = \mathbf{P}(Y = j)$; we have

$$v_j = q\mathbf{P}(X = j) + \sum_{k \neq j} p\mathbf{P}(X = k) = qu_j + p(1 - u_j)$$

so that

$$H(Y) = -\sum_{j=1}^{d} v_j \log(v_j) = \sum_{j=1}^{d} \lambda(p + (1 - dp)u_j),$$

where $\lambda(x) = -x \log(x)$. This gives the formula

$$I(X, Y) = H(Y) - H(Y \mid X)$$
$$= \sum_{j=1}^{d} \lambda(p + (1 - dp)u_j) + q \log(q) + (d - 1)p \log(p),$$

an expression which is maximal when X is uniform ($u_j = 1/d$ for all j), so that Y is uniform as well and $H(Y) = \log(d)$. Consequently,

$$I(C) = \log(d) + q \log(q) + (d - 1)p \log(p).$$

Let us now take the cost condition into account. If c_j is the cost of symbol j, we have

$$\mathbf{E}(c(X)) = \sum_{j=1}^{d} c_j u_j.$$

In the case where $(\sum_{j=1}^{d} c_j)/d \leq \gamma$, the solution $u_j = 1/d$ is admissible, and we have

$$I(C, \gamma) = I(C).$$

On the other hand, if $c_j > \gamma$ for all j, the cost condition is never satisfied and we have $I(C, \gamma) = 0$.

It remains to treat the intermediate case, when $\inf(c_j) \leq \gamma < (\sum c_j)/d$, and it does not seem possible to give an explicit formula. The question amounts to maximizing the function

4.2. Coding.

$$(u_1, \ldots, u_d) \mapsto \sum_{j=1}^{d} \lambda(p + (1-dp)u_j)$$

on a convex compact polytope of $\mathbf{R}^d$ defined by the relations

$$u_1, \ldots, u_d \geq 0, \quad \sum_{j=1}^{d} u_j = 1, \quad \sum_{j=1}^{d} c_j u_j \leq \gamma.$$

We know that this maximum is reached on a convex compact subset Λ of this polytope. We'll see at the end that one may neglect the positivity conditions on the u_j; then the method of Lagrange multipliers implies that for $u \in \Lambda$, the vector

$$(1-dp)\Big(\lambda'(p + (1-dp)u_1), \ldots, \lambda'(p + (1-dp)u_d)\Big)$$

is a multiple of the vector $(1, \ldots, 1)$ if $\sum c_j u_j < \gamma$, and is a linear combination of the vectors $(1, \ldots, 1)$ and $(c_1, \ldots, c_d)$ otherwise. In the first case, all u_j are equal, then $u_j = 1/d$ for all j, which is impossible in view of the assumption $(\sum c_j)/d > \gamma$. Therefore, there exists $a, b \in \mathbf{R}$ such that $\lambda'(u_j) = a + bc_j$ for all j. Since $\lambda'(x) = -1 - \log(x)$, we modify the notation and write

$$-\log(p + (1-dp)u_j) = a + bc_j.$$

This gives $p + (1-dp)u_j = e^{-a} e^{-bc_j}$ then

$$u_j = \frac{1}{1-dp} e^{-a} e^{-bc_j} - \frac{p}{1-dp}.$$

Set $F(b) = \sum_{j=1}^{d} e^{-bc_j}$. The condition $\sum u_j = 1$ implies that

$$e^{-a} F(b) = 1,$$

hence

$$u_j = \frac{1}{(1-dp)F(b)} e^{-bc_j} - \frac{p}{1-dp}. \tag{4.2.0.1}$$

Then the condition $\sum c_j u_j = \gamma$ becomes

$$\frac{1}{(1-dp)F(b)} \sum c_j e^{-bc_j} = \gamma + \frac{p}{1-dp} \sum c_j,$$

hence

$$(1-dp)\gamma + dp\overline{\gamma} = -\frac{F'(b)}{F(b)}, \tag{4.2.0.2}$$

where we have set $\overline{\gamma} = (\sum c_j)/d$, the cost that we need to allow for the uniform law on X. It remains to observe that the function $b \mapsto -F'(b)/F(b)$ is strictly increasing: its derivative is $(F'^2 - FF'')/F^2$; the numerator is

$$\left(\sum c_j e^{-bc_j}\right)^2 - \left(\sum e^{-bc_j}\right) \times \left(\sum c_j^2 e^{-bc_j}\right),$$

hence is positive, because of the Cauchy–Schwarz inequality, and only vanishes when all of the c_j are equal, which is not the case because $\inf(c_j) \leq \gamma < \overline{\gamma}$.

When $b \to 0$, the ratio $-F'(b)/F(b)$ tends to $(\sum c_j)/d = \overline{\gamma}$; while when $b \to +\infty$, it tends to $\inf(c_j)$. Since $\inf(c_j) \leq \gamma < (1-dp)\gamma + dp\overline{\gamma} < \overline{\gamma}$, there exists a unique real number $b \in \mathbf{R}_+^*$ for which equation (4.2.0.2) holds, and the relations (4.2.0.1) give the values of the u_j.

Exercise 2.14. [p. 93]   *a)* If there is no transmission error, we have $m_1' = m_1, \ldots, m_4' = m_4$, $m_5' = c_1$, $m_6' = c_2$ and $m_7' = c_3$.

Then, $e_1 = m_2 + m_3 + m_4 - c_1 = 0$; similarly, $e_2 = e_3 = 0$.

b) We assume that there is a transmission error at symbol i, but nowhere else. Then, $e_j = 1$ if m_i' intervenes in the definition of e_j, and $e_j = 0$ otherwise. This leads to the table:

i	e_1	e_2	e_3
1	0	1	1
2	1	0	1
3	1	1	1
4	1	1	0
5	1	0	0
6	0	1	0
7	0	0	1

c) Each possible error leads to a distinct (e_1, e_2, e_3), moreover distinct from $(0, 0, 0)$. Consequently, if we make the hypothesis that there was at most one transmission error, we see from each of the 8 possibilities for the triplet (e_1, e_2, e_3) whether there was an error (when $(e_1, e_2, e_3) \neq (0, 0, 0)$), and where this error takes place. (A more clever definition of this code would have led to a direct definition of i in terms of (e_1, e_2, e_3): that triplet is the binary expansion of i.)

Once we know that the error took place at symbol i, we know the expected value for $(m_1', \ldots, m_7')$, and we get the initial message $(m_1, \ldots, m_4)$.

d) Let us argue by contraposition, assuming that $m = g_\Phi(m')$. This means that the words $f_\Phi(m) = (m, c)$ and m' differ by at most one symbol, so that the channel made at most one transmission error.

e) The probability π that $m \neq g_\Phi(m')$ is exactly the probability that there were at least 2 transmission errors, and $1 - \pi$ is the probability that there was at most one transmission error. We thus have

$$1 - \pi = (1-p)^7 + \sum_{i=1}^{7}(1-p)^6 p = (1-p)^6(1+6p).$$

4.3. Sampling

Exercise 3.1. [p. 132]   *a)* The statement of the exercise does not define the function f on integer multiples of π; fixing $f(n\pi) = 0$ for all $n \in \mathbf{Z}$, we get a piecewise continuous function on $\mathbf{R}$; it is 2π-periodic and odd. In particular, its cosine Fourier coefficients vanish; let us compute the sine coefficients. For every $n \in \mathbf{N}^*$, we have

$$b_n(f) = \frac{2}{2\pi} \int_{-\pi}^{\pi} f(t) \sin(nt)\, dt = \frac{2}{\pi} \int_0^{\pi} f(t) \sin(nt)\, dt$$

$$= \frac{2}{\pi} \int_0^{\pi} \sin(nt)\, dt = \frac{2}{\pi} \left[-\frac{1}{n} \cos(nt) \right]_0^{\pi}$$

$$= \frac{2}{\pi n}(1 - \cos(n\pi)) = \begin{cases} 0 & \text{if } n \text{ is even;} \\ 4/\pi n & \text{otherwise.} \end{cases}$$

b) The function f is piecewise $\mathscr{C}^1$. We may thus apply Dirichlet's theorem: for every $t \in \mathbf{R}$, the Fourier series of f at t converges to the average of the left and right limits of f at t.

For $t = 0$, the left limit is -1, and the right limit is 1. Since the real Fourier series of f has only sine terms; once evaluated at $t = 0$, it vanishes identically and Dirichlet's theorem gives the obvious equality $0 = \frac{1}{2}(-1 + 1) = 0$.

The function f is continuous at $t = \pi/2$, with value 1. Moreover, if $n \in \mathbf{N}$ is odd, let us write $n = 2p + 1$, with $p \in \mathbf{N}$; then $\sin(n\pi/2) = (-1)^p$ and $b_n(f) = 4/\pi n = 4/(2p+1)\pi$. Dirichlet's theorem implies

$$\sum_{p=0}^{\infty} \frac{(-1)^p}{2p+1} = \frac{\pi}{4}.$$

c) The left-hand side of Parseval's theorem is

$$\frac{1}{2\pi} \int_0^{2\pi} |f(t)|^2\, dt = \frac{1}{2\pi} \int_0^{2\pi} dt = 1.$$

Since the cosine Fourier coefficients $a_n(f)$ vanish, its right-hand side is

$$\frac{1}{2} \sum_{n=1}^{\infty} |b_n(f)|^2 = \frac{1}{2} \sum_{p=0}^{\infty} \frac{16}{\pi^2(2p+1)^2}.$$

We then obtain the relation

$$\sum_{p=0}^{\infty} \frac{1}{(2p+1)^2} = \frac{\pi^2}{8}.$$

Exercise 3.2. [p. 132]   *a)* The function f is continuous, piecewise $\mathscr{C}^1$ and even; we compute its cosine Fourier coefficients. For $n \in \mathbf{N}$, we have

$$a_n(f) = \frac{2}{2\pi}\int_{-\pi}^{\pi} f(t)\cos(nt)\,dt = \frac{2}{\pi}\int_0^{\pi}(\pi - t)\cos(nt)\,dt.$$

We will integrate by parts but the case $n = 0$ needs an independent computation. We have

$$a_0(f) = \frac{2}{\pi}\int_0^{\pi}(\pi - t)\,dt = \frac{2}{\pi}\left[\pi t - \frac{1}{2}t^2\right]_0^{\pi} = \pi.$$

On the other hand, for $n \neq 0$, we have

$$a_n(f) = \frac{2}{\pi}\left[(\pi - t)\frac{1}{n}\sin(nt)\right]_0^{\pi} + \frac{2}{\pi n}\int_0^{\pi}\sin(nt)\,dt$$

$$= \frac{2}{\pi n}\left[-\frac{1}{n}\cos(nt)\right]_0^{\pi}$$

$$= \frac{2}{\pi n^2}(1 - \cos(n\pi)) = \begin{cases} 0 & \text{if } n \text{ is even;} \\ 4/\pi n^2 & \text{otherwise.} \end{cases}$$

b) By Dirichlet's theorem at $t = 0$, we have

$$\pi = f(0) = \frac{1}{2}\pi + \frac{4}{\pi}\sum_{p=0}^{\infty}\frac{1}{(2p+1)^2}.$$

Consequently,

$$\sum_{p=0}^{\infty}\frac{1}{(2p+1)^2} = \frac{\pi^2}{8}.$$

We remark that the series $S = \sum_{n=1}^{\infty}\frac{1}{n^2}$ contains the preceding one as the sum of all terms with odd index, and the remainder is the sum of all terms with even index. If $n = 2m$, then $1/n^2 = 1/(4m^2)$, so that the sum of all terms of even index equals $S/4$. Therefore,

$$S = \frac{\pi^2}{8} + \frac{1}{4}S,$$

hence $3S/4 = \pi^2/8$ and then

$$\sum_{n=1}^{\infty}\frac{1}{n^2} = \frac{\pi^2}{6}.$$

The third series will be computed by applying Parseval's identity to f. First we have

$$\frac{1}{2\pi}\int_{-\pi}^{\pi} f(t)^2\,dt = \frac{1}{\pi}\int_0^{\pi}(\pi - t)^2\,dt = \frac{1}{\pi}\left[-\frac{1}{3}(\pi - t)^3\right]_0^{\pi} = \frac{\pi^2}{3},$$

4.3. Sampling.

so that

$$\frac{\pi^2}{3} = \frac{1}{4}\pi^2 + \frac{1}{2}\sum_{p=0}^{\infty}\frac{16}{\pi^2}\frac{1}{(2p+1)^4},$$

hence

$$\sum_{p=0}^{\infty}\frac{1}{(2p+1)^4} = \frac{\pi^4}{96}.$$

Finally, by the same argument as for the second one, the fourth series, say, T, can be identified with the third plus 2^{-4}T. Consequently, $\frac{15}{16}$T $= \frac{\pi^4}{96}$ and

$$\sum_{p=0}^{\infty}\frac{1}{n^4} = \frac{\pi^4}{90}.$$

Leonhard EULER (1707–1783) was the first to compute the sum of the series $\sum_{n=1}^{\infty} 1/n^2 = \pi^2/6$, in 1734, thereby solving the *Basel problem* that had been put forward by the Italian mathematician Pietro Mengoli in 1650. The city of Basel, from where this problem took its name, was the birthplace of Euler, but also of the brothers Jacob and Johann Bernoulli, who had also tried to solve the problem. In fact, Euler's 1734 solution was not completely satisfactory. With some audacity, from the fact that the sine function vanishes at all integer multiples of π, he concluded the following infinite product:

$$\sin(x) = x\prod_{n=1}^{\infty}\left(1 - \frac{x^2}{n^2\pi^2}\right).$$

The required equality followed by identifying the coefficients of x^3 in the Taylor expansion of both sides. Only in 1741 would Euler justify this proof rigorously. More generally, the Euler–MacLaurin formula allows all the sums $\sum_{n=1}^{\infty} 1/n^k$ to be computed for k an *even* integer: one gets

$$\sum_{n=1}^{\infty}\frac{1}{n^k} = (-1)^{k/2-1}(2\pi)^k \frac{1}{2 \times k!}B_k,$$

where B_k is a rational number ("Bernoulli number") that is defined by way of the Taylor series of the function $x/(e^x - 1)$. The case where the integer k is odd is much more subtle and there is no similar formula. Beyond that question, Euler's contributions to modern physics and mathematics are tremendous: number theory, graph theory, astronomy, optics, fluid mechanics...

Exercise 3.3. [p. 132]   *a)* The function f is odd, so that its cosine Fourier coefficients vanish; let us compute its sine Fourier coefficients. For any integer $n \geqslant 1$, we have

$$b_n(f) = \frac{2}{2\pi}\int_{-\pi}^{\pi} f(t)\sin(nt)\,dt = \frac{2}{\pi}\int_{0}^{\pi} t(\pi - t)\sin(nt)\,dt.$$

Let us integrate by parts; we get

$$b_n(f) = \frac{2}{\pi}\left[-\frac{1}{n}t(\pi-t)\cos(nt)\right]_0^\pi + \frac{2}{n\pi}\int_0^\pi (\pi-2t)\cos(nt)\,dt$$

$$= \frac{2}{n\pi}\left[\frac{1}{n}(\pi-2t)\sin(nt)\right]_0^\pi - \frac{2}{n^2\pi}\int_0^\pi (-2)\sin(nt)\,dt$$

$$= \frac{2}{n^2\pi}\left[-2\frac{1}{n}\cos(nt)\right]_0^\pi = \begin{cases} 0 & \text{if } n \text{ is odd;} \\ 8/n^3\pi & \text{otherwise.} \end{cases}$$

b) The function f is continuous and piecewise $\mathscr{C}^1$. By Dirichlet's theorem, its Fourier series converges to f at any point. Let us take $t = \pi/2$; writing an odd integer n as $n = 2p+1$, we have $\sin(nt) = (-1)^p$, so that we get

$$f(\pi/2) = \frac{\pi^2}{4} = \sum_{p=0}^{\infty} \frac{8}{(2p+1)^3\pi}(-1)^p,$$

and finally

$$\sum_{n=0}^{\infty} \frac{(-1)^n}{(2n+1)^3} = \frac{\pi^3}{32}.$$

Let us now apply the Parseval theorem. It writes as

$$\sum_{p=0}^{\infty}\frac{1}{(2p+1)^6} = \frac{\pi^2}{64}\sum_{n=1}^{\infty} b_n(f)^2 = \frac{\pi}{64}\int_{-\pi}^{\pi} f(t)^2\,dt = \frac{\pi}{32}\int_0^\pi (t(\pi-t))^2\,dt.$$

Let us compute this integral:

$$\int_0^\pi t^2(\pi-t)^2\,dt = \int_0^\pi \left(\pi^2 t^2 - 2\pi t^3 + t^4\right)dt$$

$$= \pi^2 \times \frac{1}{3}\pi^3 - 2\pi \times \frac{1}{4}\pi^4 + \frac{1}{5}\pi^5 = \left(\frac{1}{3}-\frac{1}{2}+\frac{1}{5}\right)\pi^5 = \frac{1}{30}\pi^5.$$

Finally, we have

$$\sum_{p=0}^{\infty} \frac{1}{(2p+1)^6} = \frac{\pi^6}{960}.$$

To compute the sum $S = \sum_{n=1}^{\infty} \frac{1}{n^6}$, we proceed as in the previous exercises, separating the sum into two sums, respectively over the odd and the even indices. The contribution of the former is $\pi^4/960$, as we just computed. Writing $n = 2m$, we see that the contribution of the latter is $S/2^6 = S/64$. Consequently,

$$S\frac{63}{64} = \frac{\pi^6}{960},$$

hence

4.3. Sampling.

$$S = \frac{64}{63 \times 960}\pi^6 = \frac{1}{15 \times 63}\pi^6 = \frac{\pi^6}{945}.$$

Exercise 3.4. [p. 133]   a) The statement of the exercise does not specify f at the integer multiples of π, but if we extend it by $f(n\pi) = \cos(\pi a)$ for $n \in \mathbf{Z}$, we obtain a continuous function, piecewise $\mathscr{C}^1$, 2π-periodic and even. Its sine Fourier coefficients vanish; let us compute the cosine coefficients. The case where $a = m \in \mathbf{Z}$ is obvious: f is then a trigonometric polynomial, hence we get $a_n(f) = 0$ for $n \ne m$ and $a_m(f) = 1$. Let us now assume that $a \notin \mathbf{Z}$.

For $n \in \mathbf{N}$, we have

$$a_n(f) = \frac{2}{2\pi}\int_{-\pi}^{\pi} f(t)\cos(nt)\,dt = \frac{2}{\pi}\int_0^{\pi} \cos(at)\cos(nt)\,dt.$$

To compute this integral, we "linearize" the product of cosines:

$$\cos(at)\cos(nt) = \frac{1}{2}\Big(\cos\big((n+a)t\big) + \cos\big((n-a)t\big)\Big),$$

so that

$$a_n(f) = \frac{1}{\pi}\int_0^{\pi} \cos\big((n+a)t\big)\,dt + \frac{1}{2\pi}\int_0^{\pi} \cos\big((n-a)t\big)\,dt.$$

Since a is not an integer, we have $n + a \ne 0$, hence

$$\int_0^{\pi} \cos\big((n+a)t\big)\,dt = \left[\frac{1}{n+a}\sin\big((n+a)t\big)\right]_0^{\pi}$$
$$= \frac{1}{n+a}\sin\big((n+a)\pi\big)$$
$$= \frac{\sin(\pi a)}{n+a}(-1)^n.$$

Applying this formula for a and $-a$, we get

$$a_n(f) = (-1)^n\frac{\sin(\pi a)}{\pi}\left(\frac{1}{n+a} - \frac{1}{n-a}\right) = (-1)^n\frac{\sin(\pi a)}{\pi} \times \frac{2a}{a^2 - n^2}.$$

b) The 2π-periodic function f being continuous and piecewise $\mathscr{C}^1$, Dirichlet's theorem states the relation

$$\cos(at) = \frac{1}{2}a_0(f) + \sum_{n=1}^{\infty} a_n(f)\cos(nt)$$
$$= \frac{\sin(\pi a)}{\pi}\left(\frac{1}{a} + \sum_{n=1}^{\infty}\frac{2a}{a^2 - n^2}(-1)^n\cos(nt)\right),$$

for all $t \in [-\pi;\pi]$.

Set $t = \pi$ and divide both sides of this equality by $\sin(\pi a)/\pi$. We obtain

$$\pi \cot(\pi a) = \frac{1}{a} + \sum_{n=1}^{\infty} \frac{2a}{a^2 - n^2},$$

as required.

c) Using the asymptotic expansion $\log(1 - a^2/n^2) = O(1/n^2)$, we observe that the infinite product of the right-hand side converges to a real number which is non-zero for all $a \notin \mathbf{Z}$, and defines a differentiable function of a. Actually, the logarithmic derivative of the required relation is that of the preceding question, so that it holds up to a multiplicative factor on each interval of the form $]m; m+1[$. When a tends to 0, the asymptotic expansion $\sin(\pi a) \sim \pi a$ shows that the required equality holds on $]-1; 1[$. In fact, the right-hand side is a 2-periodic function of a. More precisely, the following computations show that when a is changed into $a + 1$, the right-hand side is changed into its opposite:

$$\pi(a+1) \prod_{n=1}^{\infty} \left(1 - \frac{(a+1)^2}{n^2}\right)$$

$$= \lim_{N \to +\infty} \pi(a+1) \prod_{n=1}^{N} \left(1 - \frac{a+1}{n}\right) \times \left(1 + \frac{a+1}{n}\right)$$

$$= \lim_{N \to +\infty} \pi(a+1) \prod_{n=1}^{N} \left(\frac{n-1}{n} - \frac{a}{n}\right) \times \left(\frac{n+1}{n} + \frac{a}{n}\right)$$

$$= \lim_{N \to +\infty} \pi(a+1)(-a) \prod_{n=2}^{N} \frac{n-1}{n}\left(1 - \frac{a}{n-1}\right) \times \frac{n+1}{n}\left(1 + \frac{a}{n+1}\right)$$

$$= \lim_{N \to +\infty} -\pi a(a+1) \frac{N+1}{N} \prod_{n=1}^{N-1}\left(1 - \frac{a}{n}\right) \prod_{n=2}^{N+1}\left(1 + \frac{a}{n}\right)$$

$$= \lim_{N \to +\infty} -\pi a \frac{N+1}{N}\left(1 - \frac{a}{N}\right)^{-1}\left(1 + \frac{a}{N+1}\right) \times \prod_{n=1}^{N}\left(1 - \frac{a}{n}\right) \prod_{n=1}^{N}\left(1 + \frac{a}{n}\right)$$

$$= -\pi a \prod_{n=1}^{\infty}\left(1 - \frac{a^2}{n^2}\right).$$

Since the function $\sin(\pi a)$ satisfies the same functional equation, we get the equality for any $a \in \mathbf{R}$.

4.3. Sampling.

Exercise 3.5. [p. 133]   a) Let $n \in \mathbf{Z}$. By definition, we have

$$c_n(f) = \frac{1}{2\pi} \int_0^{2\pi} f(x) e^{-inx} \, dx = \frac{1}{2\pi} \int_0^{2\pi} e^{px} e^{-inx} \, dx$$

$$= \frac{1}{2\pi} \int_0^{2\pi} e^{(p-in)x} \, dx = \frac{1}{2\pi} \left[\frac{1}{p-in} e^{(p-in)x} \right]_0^{2\pi}$$

$$= \frac{e^{2\pi p} - 1}{2\pi (p-in)}$$

$$= (p+in) \frac{e^{2\pi p} - 1}{2\pi (p^2 + n^2)},$$

where we used that $p \neq in$ since $p \in \mathbf{R}^*$.

b) The Fourier series of f writes as $S(f)(x) = \sum c_n(f) e^{inx}$. The function f is piecewise $\mathscr{C}^1$; we may thus apply Dirichlet's theorem: for any $x \in \mathbf{R}$, we have

$$S_N(f)(x) = \sum_{n=-N}^{N} c_n(f) e^{inx} \xrightarrow[N \to \infty]{} \frac{1}{2} \left(f(x^-) + f(x^+) \right).$$

Let us take $x = 0$. The function f tends to 1 at 0^+ and to $e^{2\pi}$ at $2\pi^-$, hence to $e^{2\pi}$ at 0^-, so that $S_N(f)(0)$ converges to $\frac{1}{2}(e^{2\pi p} + 1)$. In $S_N(f)(0)$, let us combine the terms with indices n and $-n$:

$$S_N(f)(0) = \frac{e^{2\pi p} - 1}{2\pi p} + \sum_{n=1}^{N} \frac{2p(e^{2\pi p} - 1)}{2\pi (p^2 + n^2)}$$

$$= \frac{e^{2\pi p} - 1}{2\pi p} + \frac{2p(e^{2\pi p} - 1)}{2\pi} \sum_{n=1}^{N} \frac{1}{p^2 + n^2}.$$

Setting $S = \sum_{n=0}^{\infty} 1/(p^2 + n^2)$ and dividing both sides by $(e^{2\pi p} - 1)/2\pi$, we thus obtain,

$$\frac{1}{p} + 2p \left(S - \frac{1}{p^2} \right) = \pi \frac{e^{2\pi p} + 1}{e^{2\pi p} - 1} = \pi \coth(\pi p),$$

hence

$$S = \frac{1}{p^2} - \frac{1}{2p} + \frac{\pi}{2} \coth(\pi p).$$

Exercise 3.6. [p. 133]   a) Since g is T-periodic, we have

$$f * g(x + T) = \frac{1}{T} \int_0^T f(t) g(x + T - t) \, dt = \frac{1}{T} \int_0^T f(t) g(x - t) \, dt = f * g(x),$$

which proves that $f * g$ is T-periodic.

b) Let us prove that $f * g$ is continuous. Let G be a "primitive" of g, defined by $G(x) = \int_0^x g(t) \, dt$. Since g is piecewise continuous, the function G is continuous and piecewise $\mathscr{C}^1$; it admits right and left derivatives at each point. We first treat the case where f is a step function, that is, we assume that there exists a subdivision $(t_0, \ldots, t_n)$ of $[0; T]$ such that for any $i \in \{1, \ldots, n\}$, the restriction of f to the open interval $]t_{i-1}; t_i[$ is a constant c_i. Then we have

$$f * g(x) = \frac{1}{T} \sum_{i=1}^{n} c_i \int_{t_{i-1}}^{t_i} g(x-t)\,dt = \frac{1}{T} \sum_{i=1}^{n} c_i \big(G(x-t_{i-1}) - G(x-t_i)\big),$$

and this formula implies that $f * g$ is continuous, even left and right differentiable at each point.

In the general case, we use the fact that we can approximate f uniformly by step functions $\tilde{f}$; this gives a uniform approximation of $f * g$ by the continuous function $\tilde{f} * g$, and this implies that $f * g$ is continuous.

c) Let $n \in \mathbf{Z}$. We write

$$c_n(f * g) = \frac{1}{T} \int_0^T (f * g)(x) e^{-2i\pi nx/T}\,dx$$

$$= \frac{1}{T^2} \int_0^T \left(\int_0^T f(t) g(x-t)\,dt \right) e^{-2i\pi nx/T}\,dx.$$

Applying Fubini's theorem, we exchange the order of integration and write

$$c_n(f * g) = \frac{1}{T^2} \int_0^T \left(f(t) g(x-t) e^{-2i\pi nx/T}\,dx \right) dt.$$

In the inner integral, t is fixed, so that this integral equals

$$f(t) \int_0^T g(x-t) e^{-2i\pi nx/T}\,dx.$$

Let us now make the change of variables $y = x - t$; we get

$$f(t) \int_{x-T}^{x} g(y) e^{-2i\pi n(y+t)/T}\,dy = f(t) e^{-2i\pi nt/T} \int_{x-T}^{x} g(y) e^{-2i\pi ny/T}\,dy.$$

In this expression, we recognize the integral that defines the Fourier coefficient $c_n(g)$, but on the interval $[x - T; x]$; since this is a period, this integral is equal to $Tc_n(g)$. We then have

$$c_n(f * g) = \frac{1}{T} \int_0^T f(t) e^{-2i\pi nt/T} c_n(g)\,dt = c_n(f) \times c_n(g).$$

Exercise 3.7. [p. 133]   a) The function F is continuous at any point t such that $t - a$ is not a multiple of T. As $t \to a$, $F(t)$ tends to $f'(a)/(2i\pi/T)$ (L'Hôpital's rule!), which proves that F can be extended to a continuous function.

Let $\mathbf{1}$ be the constant function with value 1 and let ε be the function given by $\varepsilon(t) = e^{2i\pi t/T}$. The relation $f(t) = f(a) + (e^{2i\pi(t-a)/T} - 1)F(t)$ can be written

4.3. Sampling.

$$f = f(a)\mathbf{1} + e^{-2i\pi a/T}\varepsilon F - F,$$

so that we have the following relation

$$c_n(f) = f(a)c_n(\mathbf{1}) + e^{-2i\pi a/T}c_n(\varepsilon F) - c_n(F)$$

for its Fourier coefficients. We observe that

$$c_n(\varepsilon F) = \frac{1}{T}\int_0^T \varepsilon(t)F(t)e^{-2i\pi nt/T}\,dt = \frac{1}{T}\int_0^T F(t)e^{-2i\pi(n-1)t/T}\,dt = c_{n-1}(F).$$

On the other hand, we have $c_n(\mathbf{1}) = 1$ if $n = 0$ and $c_n(\mathbf{1}) = 0$ otherwise, so that $c_n(\mathbf{1}) = \delta_n$ (Kronecker function). Finally,

$$c_n(f) = f(a)\delta_n + e^{-2i\pi a/T}c_{n-1}(F) - c_n(F).$$

b) Let us take $t = a$. When we sum the Fourier series from $M < 0$ to $N > 0$, we obtain

$$\sum_{n=M}^{N} c_n(f)e^{2i\pi nc/T} = f(a) + \sum_{n=M}^{N}\left(e^{-2i\pi a/T}c_{n-1}(F) - c_n(F)\right)e^{2i\pi nc/T}$$

$$= f(a) + \sum_{n=M}^{N}\left(c_{n-1}(F)e^{2i\pi(n-1)a/T} - c_n(F)e^{2i\pi nc/T}\right)$$

$$= f(a) + c_{M-1}(F)e^{2i\pi(M-1)a/T} - c_N(F)e^{2i\pi Nc/T}.$$

Since F is continuous, it is locally integrable and its Fourier coefficients converge to 0 at $\pm\infty$. This implies the convergence

$$\lim_{\substack{M\to-\infty \\ N\to+\infty}} \sum_{n=M}^{N} c_n(f)e^{2i\pi nc/T} = f(a).$$

The same proof works under the mere hypothesis that F is locally integrable, because it is sufficient to imply the existence of the Fourier coefficients of F, as well as their convergence to 0 at infinity. In particular, it is enough to assume that f is locally integrable, and has left and right derivatives at $t = a$.

c) We write

$$c_n(\varphi) = \frac{1}{T}\int_{a-T/2}^{a+T/2}\varphi(t)e^{-2i\pi nt/T}\,dt$$

$$= \frac{1}{T}\left(-\int_{a-\delta}^{a}e^{-2i\pi nt/T}\,dt + \int_{a}^{a+\delta}e^{-2i\pi nt/T}\,dt\right)$$

$$= \frac{1}{T}e^{-2i\pi na/T}\left(-\int_{-\delta}^{0}e^{-2i\pi nt/T}\,dt + \int_{0}^{\delta}e^{-2i\pi nt/T}\,dt\right)$$

$$= \frac{1}{T} e^{-2i\pi na/T} 2i \int_0^\delta \sin(2\pi nt/T) \, dt.$$

In particular, we have $c_0(\varphi) = 0$. For $n \neq 0$, we get

$$c_n(\varphi) = \frac{2i}{T} e^{-2i\pi na/T} \left[-\frac{T}{2\pi n} \cos(2\pi nt/T) \right]_0^\delta = \frac{i}{\pi n} e^{-2i\pi na/T} (1 - \cos(2\pi n\delta/T)).$$

At $t = a$, the Fourier series of φ is given by

$$\sum_{n \neq 0} \frac{i}{\pi n} (1 - \cos(2\pi n\delta/T)).$$

When summed symmetrically, the terms with indices $-n$ and n cancel. Consequently, the Fourier series converges to 0 at $t = a$.

We also observe that φ has right and left limits at a, namely $\varphi(a^+) = 1$ and $\varphi(a^-) = -1$, with average 0.

d) Let us consider the function g given by

$$g(t) = f(t) - \frac{1}{2}(f(a^+) - f(a^-))\varphi(t).$$

It is piecewise $\mathscr{C}^1$. When t tends to a (on the left, or on the right), it converges to $\frac{1}{2}(f(a^+) + f(a^-))$; Let us extend it by continuity at a. Then the exercise applies to g, hence the Fourier series of g converges to $g(a)$ at $t = a$.

The Fourier series of f is that of g plus that of φ, multiplied by $(f(a^+) - f(a^-))/2$. The preceding question implies that it converges to $g(a)$ at $t = a$.

Since $g(a) = \frac{1}{2}(f(a^+) + f(a^-))$, this concludes the exercise.

Exercise 3.8. [p. 134]   a) Let f be the 2π-periodic function such that $f(0) = f(\pi) = 0$, $f(x) = 1$ for $0 < x < \pi$, and $f(x) = -1$ for $-\pi < x < 0$. It is odd and piecewise $\mathscr{C}^1$. Its sine Fourier series is given by

$$\sum_{n=1}^\infty b_n(f) \sin(nx),$$

where

$$b_n(f) = \frac{2}{2\pi} \int_{-\pi}^\pi f(x) \sin(nx) \, dx$$

$$= \frac{2}{\pi} \int_0^\pi \sin(nx) \, dx$$

$$= \frac{2}{\pi} \left[-\frac{1}{n} \cos(nx) \right]_0^\pi = \begin{cases} 4/\pi n & \text{if } n \text{ is odd}; \\ 0 & \text{otherwise}. \end{cases}$$

4.3. Sampling.

In other words, $S_n(x)$ is the $(2n+1)$th partial sum of the Fourier series of $\frac{\pi}{4}f$. Let $x \in \,]0;\pi[$. Since f is piecewise $\mathscr{C}^1$ and is continuous at x, Dirichlet's theorem implies that $S_n(x)$ converges to $\frac{\pi}{4}f(x) = 4/\pi$.

b) We have $S'_n(x) = \sum_{k=1}^{n} \cos\bigl((2k-1)x\bigr)$. To simplify this sum, we write $\cos\bigl((2k-1)x\bigr) = \operatorname{Re}(e^{i(2k-1)x})$, so that $S'_n(x)$ is the real part of

$$\sum_{k=1}^{n} e^{i(2k-1)x} = e^{ix}\frac{e^{2inx}-1}{e^{2ix}-1},$$

the sum of n terms of a geometric progression with first term e^{ix} and common ratio e^{2ix}. This expression simplifies as

$$e^{ix}\frac{e^{inx}(2i\sin(nx))}{e^{ix}(2i\sin(x))} = e^{inx}\frac{\sin(nx)}{\sin(x)},$$

so that

$$S'_n(x) = \frac{\cos(nx)\sin(nx)}{\sin(x)} = \frac{\sin(2nx)}{2\sin(x)}.$$

This computation is only valid when $e^{2ix} \neq 1$, that is, if x is not an integer multiple of π. In that case, we see directly that $S'_n(x) = n$ for $x \equiv 0 \pmod{2\pi}$ and $S'_n(x) = -n$ for $x \equiv \pi \pmod{2\pi}$.

Since S_n is 2π-periodic and odd, we study its variations on the interval $[0;\pi]$. For $x \in \,]0;\pi[$, the denominator $2\sin(x)$ is strictly positive, but the numerator $\sin(2nx)$ vanishes and changes sign at the points x_k given by $x_k = k\pi/2n$, for $k \in \{0,\ldots,2n\}$. We conclude that S_n is increasing on $[0;x_1]$, decreasing on $[x_1;x_2]$, increasing on $[x_2;x_3]$, etc.

c) The variations of the function S_n indicate that S_n has a local maximum at each x_{2k-1}, for $k \in \{1,\ldots,n\}$, with value $m_{2k-1} = S_n(x_{2k-1}) = S_n\bigl((2k-1)\pi/2n\bigr)$. To prove the inequalities $m_1 \geqslant m_3 \geqslant \ldots$, we write, for each $k \in \{1,\ldots,n-1\}$,

$$m_{2k-1} - m_{2k+1} = S_n(x_{2k-1}) - S_n(x_{2k+1})$$
$$= -\int_{x_{2k-1}}^{x_{2k+1}} S'_n(x)\,dx$$
$$= -\int_{x_{2k-1}}^{x_{2k+1}} \frac{\sin(2nx)}{2\sin(x)}\,dx.$$

Let us make the change of variables $x = t + x_{2k} = t + k\pi/n$; since $2nx_{2k} = 2k\pi$, we get $\sin(2nx) = \sin(2nt)$ and

$$m_{2k-1} - m_{2k+1} = -\int_{-\pi/2n}^{\pi/2n} \frac{\sin(2nt)}{2\sin(x)}\,dt$$
$$= \int_0^{\pi/2n} \sin(2nt)\left(+\frac{1}{2\sin(x')} - \frac{1}{2\sin(x)}\right)dt,$$

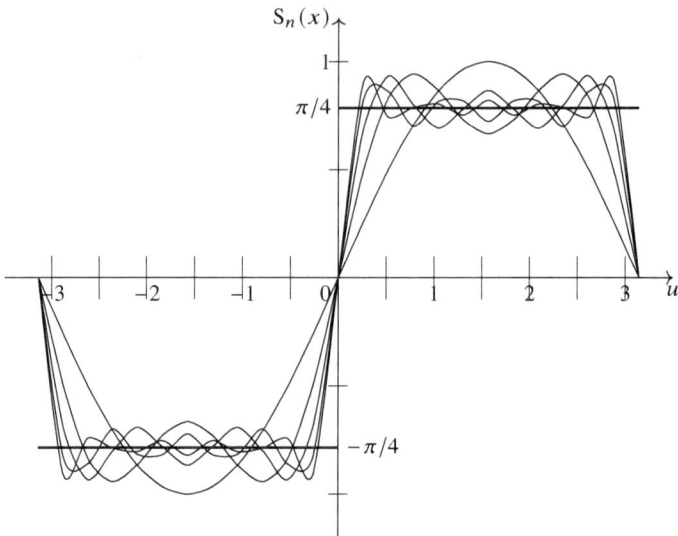

Fig. 4.2 Graph of $S_n(x)$ for $1 \leq n \leq 5$

where $x' = -t + x_{2k}$. In the interval we consider, we have $2nt \in [0; \pi]$, so that $\sin(2nt) \geq 0$. We also have

$$0 \leq x_{2k-1} \leq x' \leq x_{2k} \leq x \leq x_{2k+1} \leq \frac{1}{2}\pi,$$

so that $0 \leq \sin(x') \leq \sin(x)$, and this implies the inequality $m_{2k-1} \geq m_{2k+1}$.

d) Observe the symmetry $S_n(\pi - x) = S_n(x)$, because the nonzero sine Fourier coefficients of S_n have odd index. By the preceding question, the maximum of S_n on $[0; \pi/2]$ is reached at x_1; on $[\pi/2; \pi]$, it is reached at x_{2n-1}. Consequently,

$$\sup(S_n) = m_1 = S_n(x_1) = \int_0^{x_1} S'_n(x)\,dx = \int_0^{\pi/2n} \frac{\sin(2nx)}{2\sin(x)}\,dx.$$

Let us make the change of variables $t = 2nx$; then

$$\sup(S_n) = \int_0^\pi \frac{\sin(t)}{4n\sin(t/2n)}\,dt.$$

When $n \to +\infty$, the integrated function converges to $\sin(t)/2t$, even uniformly, so that

$$\lim_{n \to +\infty} \sup(S_n) = \frac{1}{2}\int_0^\pi \frac{\sin(t)}{t}\,dt \approx 0.926.$$

A *paradox* that can be observed from the preceding figure is that this limit is strictly greater than $\sup(f) = \pi/4 \approx 0.785$, the ratio between these two quantities is ≈ 1.179. Since the function f is not continuous, it cannot be the *uniform* limit of its Fourier series; there is even an overshoot of roughly 9% of the amplitude of discontinuity (here, $f(0^+) - f(0^-) = \pi/2$).

This is a general phenomenon. Indeed, let us consider a function u, 2π-periodic and piecewise $\mathscr{C}^1$, and assume that it is discontinuous at a point a. We can then write it as the sum of a continuous and piecewise $\mathscr{C}^1$ function, and of a linear combinations of functions which are analogous to the function f of this exercise, "placed" at each discontinuity, and whose role is to suppress that discontinuity. By itself, the function f is not sufficient because it has a discontinuity not only at 0 but also at π, and we have to consider the 2π-periodic function g given by $g(0) = 0$ and $g(t) = \pi - t$ for $t \in \,]0; 2\pi[$; however, its study is slightly more delicate.

Anyway, the Fourier series of our initial function u gives rise to local overshoots at each discontinuity, of amplitude approximately equal to 9% of the amplitude of the discontinuity.

This phenomenon was discovered by Henry Wilbraham in 1848; nowadays, it bears the name of the *Gibbs phenomenon*, after the mathematician J. Willard Gibbs, who rediscovered it in 1899. I recommend the paper of HEWITT & HEWITT (1979) for a detailed discussion of this phenomenon and of the history of its discovery. From the point of view of signal theory, it explains the ringing artifacts of a rendered signal when its source has discontinuities.

Exercise 3.9. [p. 134]   *a)* By differentiation under the integral sign, we have

$$g'(y) = \int_\mathbf{R} -2i\pi x f(x) e^{-2i\pi xy} \, dx$$
$$= i \int_\mathbf{R} (-2x f(x)) e^{-2i\pi xy} \, dx$$
$$= i \int_\mathbf{R} f'(x) e^{-2i\pi xy} \, dx$$
$$= i \left[f(x) e^{-2i\pi xy} \right]_{-\infty}^{\infty} - i \int_\mathbf{R} f(x)(-2i\pi y) e^{-2i\pi xy} \, dx$$
$$= -2\pi y \int_\mathbf{R} f(x) e^{-2i\pi xy} \, dx$$
$$= -2\pi y g(y).$$

We could also have started from the relation $f'(x) = -2\pi x f(x)$ and computed the Fourier transform of both sides. For the left-hand side, we know that

$$\widehat{f'}(y) = 2i\pi y \widehat{f}(y) = 2i\pi y g(y),$$

while the Fourier transform of the right-hand side is equal to

$$-i\mathscr{F}(-2i\pi x f(x))(y) = -i\mathscr{F}(f)'(y) = -ig'(y).$$

We then reprove the equality $g'(y) = -2\pi y g(y)$.

The general solution of this differential equation is

$$g(y) = c e^{-\pi y^2},$$

where c is any real number.

b) Let us apply Fourier transformation to both sides of the equation $g = cf$: the left-hand side becomes $\widehat{g}(y) = \widehat{\widehat{f}}(x) = f(-x)$, while the right-hand side gives $c\widehat{f}(x) = cg(x) = c^2 f(x)$, hence the equality $c^2 = 1$.

We also have $\int_\mathbb{R} f(x)\,dx = g(0) = cf(0) = c$. In particular, $c > 0$, so that $c = 1$ and $\int_\mathbb{R} e^{-\pi x^2}\,dx = 1$.

c) The Poisson summation formula (theorem 3.6.4) can be written

$$\theta(t) = \sum_{n=-\infty}^{\infty} f(\sqrt{t}n) = \frac{1}{\sqrt{t}} \sum_{n=-\infty}^{\infty} \widehat{f}(n/\sqrt{t}) = \theta(1/t)/\sqrt{t},$$

and this implies the functional equation $\theta(1/t) = \sqrt{t}\,\theta(t)$.

Exercise 3.10. [p. 135]   a) For $y \in \mathbb{R}$, one has

$$\widehat{f}(y) = \int_{-\infty}^{\infty} e^{-|x|} e^{-2i\pi xy}\,dx = \int_0^{\infty} e^{-x(1+2i\pi y)}\,dx + \int_0^{\infty} e^{-x(1-2i\pi y)}\,dx$$

$$= \frac{1}{1+2i\pi y} + \frac{1}{1-2i\pi y} = \frac{2}{1+4\pi^2 y^2}.$$

b) The Fourier inversion formula writes as

$$f(x) = e^{-|x|} = \int_\mathbb{R} \frac{2}{1+4\pi^2 y^2} e^{2i\pi xy}\,dy.$$

In particular, for $x = 0$, we get

$$1 = \int_{-\infty}^{\infty} \frac{2}{1+4\pi^2 y^2}\,dy.$$

Since the derivative of $\arctan(2\pi y)$ is $2\pi/(1+4\pi^2 y^2)$, the right-hand side can also be computed as

$$\frac{2}{2\pi}\left[\arctan(2\pi y)\right]_{-\infty}^{\infty} = \frac{2}{2\pi}\left(\frac{\pi}{2} - \frac{-\pi}{2}\right) = 1.$$

c) Plancherel's theorem states that $\int_{-\infty}^{\infty} f(x)^2\,dx = \int_{-\infty}^{\infty} \widehat{f}(y)^2\,dy$. The left-hand side is equal to

$$\int_{-\infty}^{\infty} e^{-2|x|}\,dx = 2\int_0^{\infty} e^{-2x}\,dx = 2\frac{1}{2} = 1.$$

4.3. Sampling.

Consequently,
$$\int_{-\infty}^{\infty} \frac{4}{(1+4\pi^2 y^2)^2} \, dy = 1.$$

Exercise 3.11. [p. 135]   *a)* We have

$$\widehat{f}(y) = \int_{-\infty}^{\infty} f(x)e^{-2i\pi xy} \, dx = \int_{-1}^{1} (1-|x|)e^{-2i\pi xy} \, dx$$
$$= \int_{0}^{1} (1-x) 2\cos(2\pi xy) \, dx.$$

For $y = 0$, we get
$$\widehat{f}(0) = \int_{0}^{1} 2(1-x) \, dx = 1.$$

Assume that $y \neq 0$; we can then integrate by parts, and we obtain

$$\widehat{f}(y) = \left[(1-x)\frac{2}{2\pi y} \sin(2\pi xy) \right]_0^1 + \frac{1}{\pi y} \int_0^1 \sin(2\pi xy) \, dx$$
$$= \frac{1}{\pi y} \left[-\frac{1}{2\pi y} \cos(2\pi y) \right]_0^1 = \frac{1-\cos(2\pi y)}{2\pi^2 y^2} = \frac{\sin^2(\pi y)}{\pi^2 y^2}$$
$$= \left(\operatorname{sinc}(\pi y) \right)^2.$$

b) The function $\widehat{f}$ is continuous on **R**; it decays as $1/y^2$ at infinity, hence it is integrable. We may thus apply the Fourier inversion theorem. This gives

$$f(0) = 1 = \int_{-\infty}^{\infty} \widehat{f}(y) \, dy,$$

hence the desired equality.

Exercise 3.12. [p. 135]   *a)* By definition, we have

$$\widehat{f}(y) = \int_{-\infty}^{\infty} f(x)e^{-2i\pi xy} \, dx = \int_{-1}^{1} e^{-2i\pi xy} \, dx.$$

For $y = 0$, we get $\widehat{f}(0) = 2$. Assume that $y \neq 0$; then we have

$$\widehat{f}(y) = \frac{1}{-2i\pi y} \left[e^{-2i\pi xy} \right]_{-1}^{1}$$
$$= \frac{1}{-2i\pi y} \left(e^{-2i\pi y} - e^{2i\pi y} \right)$$
$$= \frac{1}{-2i\pi y} (-2i \sin(2\pi y)) = 2\frac{\sin(2\pi y)}{2\pi y}.$$

b) Let us apply Plancherel's theorem. On the one side, we have

$$\int_{-\infty}^{\infty} |f(x)|^2 \, dx = \int_{-1}^{1} dx = 2,$$

and on the other side,

$$\int_{-\infty}^{\infty} |\widehat{f}(y)|^2 \, dy = 4 \int_{-\infty}^{\infty} \left(\frac{\sin(2\pi y)}{2\pi y} \right)^2 dy = \frac{2}{\pi} \int_{-\infty}^{\infty} \left(\frac{\sin(t)}{t} \right)^2 dt.$$

Consequently,

$$\int_{0}^{\infty} \left(\frac{\sin(t)}{t} \right)^2 dt = \frac{1}{2} \int_{-\infty}^{\infty} \left(\frac{\sin(t)}{t} \right)^2 dt = \frac{\pi}{2}.$$

c) The function $t \mapsto \sin(t)/t$ is continuous on $\mathbf{R}$; we may integrate by parts between 0 and T, taking $1 - \cos(t)$ as a primitive of $\sin(t)$. This gives

$$\int_0^T \frac{\sin(t)}{t} \, dt = \left[\frac{1 - \cos(t)}{t} \right]_0^T + \int_0^T \frac{1 - \cos(t)}{t^2} \, dt$$

$$= \frac{1 - \cos(T)}{T} + \int_0^T \frac{2 \sin^2(t/2)}{t^2} \, dt$$

$$= \frac{1 - \cos(T)}{T} + \int_0^T \frac{\sin^2(t/2)}{t^2/4} \, d(t/2)$$

$$= \frac{1 - \cos(T)}{T} + \int_0^{T/2} \frac{\sin^2(u)}{u^2} \, du.$$

When $T \to +\infty$, the left-hand side tends to 0 and the second tends to $\int_0^\infty (\sin(u)/u)^2 \, du$. This proves that the given integral converges. It also proves that

$$\int_0^{+\infty} \frac{\sin(t)}{t} \, dt = \int_0^{+\infty} \left(\frac{\sin(t)}{t} \right)^2 dt = \frac{\pi}{2}.$$

References

M. Aigner & G. M. Ziegler (2018), *Proofs from THE BOOK*, Springer Berlin Heidelberg, Berlin, Heidelberg, sixth edition.

N. Alon & J. H. Spencer (2008), *The Probabilistic Method*, Wiley-Interscience Series in Discrete Mathematics and Optimization, Wiley, Hoboken, N.J, third edition.

R. B. Blackman, H. W. Bode & C. E. Shannon (1946), "Data smoothing and prediction in fire-control systems". Summary Technical Report, Div. 7, National Defense Research Committee.

P. R. Chernoff (1980), "Pointwise Convergence of Fourier Series". *The American Mathematical Monthly*, **87** (5), p. 399.

T. M. Cover & J. A. Thomas (2006), *Elements of Information Theory*, Wiley-Interscience [John Wiley & Sons], Hoboken, NJ, second edition.

H. Dym & H. P. McKean (2016), *Séries et intégrales de Fourier*, Nouvelle bibliothèque mathématique **13**, Cassini, Paris. Translated from the 1972 English original by Éric Kouris.

R. W. Hamming (1950), "Error Detecting and Error Correcting Codes". *Bell System Technical Journal*, **29** (2), pp. 147–160.

W. Heisenberg (1927), "Über den anschaulichen Inhalt der quantentheoretischen Kinematik und Mechanik". *Zeitschrift für Physik*, **43** (3), pp. 172–198.

E. Hewitt & R. E. Hewitt (1979), "The Gibbs-Wilbraham phenomenon: An episode in Fourier analysis". *Archive for History of Exact Sciences*, **21** (2), pp. 129–160.

E. H. Kennard (1927), "Zur Quantenmechanik einfacher Bewegungstypen". *Zeitschrift für Physik*, **44** (4), pp. 326–352.

A. N. Kolmogorov (1956), *Foundation of the Theory of Probability*, Chelsea, New York.

H. J. Landau & H. O. Pollak (1961), "Prolate spheroidal wave functions, Fourier analysis and uncertainty — II". *The Bell System Technical Journal*, **40** (1), pp. 65–84.

R. Redheffer (1984), "Convergence of Fourier Series at a Discontinuity". *SIAM Journal on Mathematical Analysis*, **15** (5), pp. 1007–1009.

C. E. SHANNON (1948), "A mathematical theory of communication". *Bell System Tech. J.*, **27**, pp. 379–423, 623–656.

C. E. SHANNON (1949a), "Communication in the presence of noise". *Proc. I.R.E.*, **37**, pp. 10–21.

C. E. SHANNON (1949b), "Communication Theory of Secrecy Systems". *Bell System Technical Journal*, **28** (4), pp. 656–715.

C. E. SHANNON & W. WEAVER (1949), *The Mathematical Theory of Communication*, The University of Illinois Press, Urbana, Ill.

D. SLEPIAN & H. O. POLLAK (1961), "Prolate spheroidal wave functions, Fourier analysis and uncertainty — I". *The Bell System Technical Journal*, **40** (1), pp. 43–63.

S. A. VANSTONE & P. C. OORSCHOT (1989), *An Introduction to Error Correcting Codes with Applications*, Springer US, Boston, MA.

H. WEYL (1950), *The Theory of Groups and Quantum Mechanics*, Dover Books on Mathematics, Dover Publ, Mineola, NY. Translation of the German second edition, 1931.

WIKIPEDIA (2022), "Hearing range — Wikipedia, the free encyclopedia". URL https://en.wikipedia.org/w/index.php?title=Hearing_range&oldid=1101971335, accessed 11-August-2022.

J. WOLFOWITZ (1957), "The coding of messages subject to chance errors". *Illinois Journal of Mathematics*, **1** (4), pp. 591–606.

J. WOLFOWITZ (1958), "The maximum achievable length of an error correcting code". *Illinois Journal of Mathematics*, **2** (3), pp. 454–458.

Index

B

Banach fixed point theorem, 46
Banach, Stefan (1892–1945), 46
Basel problem, 191
Bayes's law, 18
Bernoulli, Daniel (1700–1782), 26
Bernoulli, Jacob (1655–1705), 25, 26
Bernoulli, Johann (1667–1748), 26
Berrou, Claude (1951–), 88
Bessel, Friedrich Wilhelm (1784–1846), 103
Bessel's inequality, 105
Bessel's relation, 103
Bienaymé, Irénée-Jules (1796–1878), 67
Bienaymé–Chebyshev inequality, 67

C

capacity–cost function of a transmission channel, 93
cardinal sine, 119
Cauchy criterion, 4
Cauchy, Augustin Louis (1789–1857), 4
Cauchy–Schwarz inequality, 16
Cesàro, Ernesto (1859–1906), 38
Cesàro lemma, 38
Chasles, Michel (1793–1880), 29
Chebyshev, Pafnuty (1821–1894), 67
Chernoff's inequality, 91
code, 56
    binary, 56
    for a memoryless channel, 81
    Hamming, 94
    Huffman, 63, 65
    optimal, 62
    prefix, 57, 60
    transmission rate, 81
    uniquely decodable, 57
convolution product, 133

D

data processing inequality, 37, 84, 165
Dirichlet, Peter Gustave Lejeune (1805–1859), 111
Dirichlet kernel, 112
Dirichlet regularization of a piecewise continuous function, 111
divergence, 32, 34

E

entropy, 25
    conditional, 29
    of a discrete random variable, 25
    statistical interpretation, 70
entropy rate
    of a stochastic process, 38
Erdős, Paul (1913–1996), 56
Euler, Leonhard (1707–1783), 25, 191
event, 9
expectancy
    conditional — of a random variable given a random variable, 20
expectation
    conditional —, 19
    of a discrete random variable, 13

F

Fano inequality, 82, 83
Fano, Robert M. (1917–2016), 82
Fejér kernel, 118

Fejér, Lipót (1880–1959), 118
Fourier, Joseph (1768–1830), 95
Fourier coefficients of a periodic function, 100
Fourier inversion formula, 121
Fourier series of a periodic function, 101
Fourier transform of an integrable function, 119
functional equation of the θ-function, 134

G

Gallager, Robert G. (1931–), 88
Gibbs phenomenon, 134, 201

H

Heisenberg inequality, 126
Heisenberg, Werner (1901–1976), 126
Huffman, David (1925–1999), 63
Huffman code, 63, 65
Huffman theorem, 65

I

independence
   conditional, 36
   of events, 18
   of random variables, 18
inequality
   Bienaymé–Chebyshev, 67
   Markov, 67

K

kernel
   Dirichlet, 112
   Fejér, 118
   Poisson, 117
Kolmogorov, Andrey (1903–1987), 1
Kraft, Leon, 58
Kraft–McMillan inequality, 58

L

law
   Bernoulli, 12
   geometric, 12, 21
   of a discrete random variable, 11
   of large numbers, 68
   of total probabilities, 18
   Poisson, 13
   stationary — of a Markov process, 42
   uniform, 12, 20
Lebesgue, Henri-Léon (1875–1941), 98

M

Markov, Andrey (1856–1922), 41
Markov chain, 40
   aperiodic, 48
   irreducible, 48
Markov inequality, 67
matrix
   of transmission probabilities of a channel, 74
   transition — of a Markov process, 41
McMillan, Brockway (1915–2016), 58
Minkowski inequality, 16
moment
   of a discrete random variable, 16
mutual information, 34
   conditional, 36
   statistical interpretation, 72

N

Napier, John, 25
Nyquist, Harry (1889–1976)., 96
Nyquist–Shannon theorem, 124

P

Parseval, Marc-Antoine (1755–1836), 109
Parseval's theorem, 109
Perron, Oskar (1880–1975), 44
Perron's theorem, 44
Pinsker's inequality, 91
Plancherel, Michel (1885–1967), 121
Plancherel's theorem, 121
Poisson law, 21
Poisson, Siméon Denis (1781–1840), 125
Poisson kernel, 117
Poisson summation formula, 125
possible values of a discrete random variable, 11
probability, 9
   conditional, 17

Q

quiver
   associated with a Markov chain, 41
   connected, 48

R

random variable

Bernoulli, 12
  certain —, 11
  discrete —, 11
  expectation, 13
  law of a discrete —, 11
  moment of a discrete —, 16
  possible values of a discrete —, 11
  uniform, 12
  variance of a discrete —, 16
Rao–Blackwell inequality, 53
Riemann, Bernhard (1826–1866), 108
Riemann–Lebesgue lemma, 115

S

sampling theorem, 123
Schwartz class, 121
Shannon's coding theorem, 58, 60
Shannon's theorem, 82
σ-algebra, 9
Slepian–Landau–Pollak theorem, 129
states of a Markov chain, 41
statistic
  sufficient, 53
statistical interpretation of entropy, 70
statistical interpretation of mutual information, 72
stochastic matrix, 41
  primitive, 44
stochastic process, 38
  homogeneous, 41
  Markov, 40
  stationary, 39, 42
summable family, 2
  sum, 2
  support, 6
sum of a summable family, 2

support of a summable family, 6

T

transition matrix of a Markov process, 41
transmission capacity of a memoryless channel, 75
transmission channel
  memoryless, 74
  symmetric, 79
  symmetric binary, 76
transmission rate
  accessible, 82
  of a code, 81
trigonometric polynomial, 102

U

universe, 9

V

variance
  conditional — of a random variable given a random variable, 20
  of a discrete random variable, 16

W

Wikipedia, vii

Y

Young's inequality, 16